Hormones and the Mind

Hormones and the Mind

A WOMAN'S GUIDE TO ENHANCING MOOD, MEMORY, AND SEXUAL VITALITY

Edward L. Klaiber, M.D.

Quill
An Imprint of HarperCollins*Publishers*

Hormones and the Mind is intended to be educational and in no way a substitute for specific advice from a physician or mental health professional. If you are feeling unstable, seek medical help at once. The author and publisher expressly disclaim responsibility for any negative effects directly or indirectly attributable to the use or application of any information contained in this book.

A hardcover edition of this book was published in 2001 by HarperCollins Publishers.

HarperCollins books may be purchased for educational, business, or sales promotional use. For information please write: Special Markets Department, HarperCollins Publishers Inc., 10 East 53rd Street, New York, NY 10022.

First Quill edition published 2002.

Designed by Nancy B. Field

The Library of Congress has catalogued the hardcover edition as follows:

Klaiber, Edward L.
Hormones and the mind : a woman's guide to enhancing mood, memory, and sexual vitality / Edward L. Klaiber.—1st ed.
p. cm.
Includes bibliographical references and index.
ISBN 0-06-019373-5
1. Hormone therapy. 2. Women—Mental health. 3. Menopause—Hormone therapy. I. Title.

RC483.5.H6 K53 2001
615'36'082—dc21 00-044868

ISBN 0-06-093187-6 (pbk.)

02 03 04 05 06 ❖/RRD 10 9 8 7 6 5 4 3 2 1

To my brothers in science,
Bill Vogel and Don Broverman

Contents

Acknowledgments ix

Introduction 1

1 *Deep Impact*
HORMONES AND THE BRAIN 9

2 *Reclaiming Emotional Balance*
DEPRESSION AND MOOD 31

3 *Bolstering Brain Power*
MEMORY AND MENTAL SHARPNESS 69

4 *Harnessing the Chemistry of Desire*
SEXUAL HEALTH 99

5 *Calming the Storms*
THE NEW PMS PRESCRIPTION 129

6 *The Gift of Consciousness*
ESTROGEN AND ALZHEIMER'S DISEASE 147

7 *The Protected Passage*
HORMONE MODULATION FOR
A HEALTHY MENOPAUSE 173

8 *The Art and Science of Hormone Modulation*
DIAGNOSIS AND TREATMENT 209

9 *The Safety Equation*
WEIGHING RISKS AND BENEFITS 229

10 *Taking Charge of Your Hormones*
THE DECISION IS YOURS 259

11 *New Medicine for the Mind*
THE FUTURE OF HORMONE MODULATION 269

References 277

Glossary 293

Index 303

Acknowledgments

A number of people deserve my most heartfelt appreciation for their irreplaceable efforts in the writing of this book. The first is my wife, Jeanne. Sharing my life with her has been my greatest privilege. It is to her unswerving support that I owe the richness of my family life and my success professionally. She has been a beacon of common sense, guiding me in all I do. My total respect and enduring love belong to her. She has made substantive contributions to the content of this book. Our children, Carrie, Steve, and Rob, and our son-in-law, Tom Lynch, have always enriched our lives with joy. It is with great pride that we follow their achievements and see another generation beginning with our two grandsons, Zachary and John.

Steven Power has been the only person who has been with me during the writing of every word in this book. Originally his role was to act as a word processor but very quickly, because of his prodigious intelligence and strict work ethic, he became far more. Even though a full-time college student, he has devoted countless hours to the necessary library and computer research. He acted as the stand-in for the reader, steering me away from my normal pedantic style of writing, urging me to be more reader friendly and alerting me to overdoses of science. Throughout this difficult and laborious process he has made suggestions regarding the text and transferred his stability to the pages of this book. He was a true contributor in every sense of the word. For that I am extremely grateful. He has become a valued colleague and my trusted friend.

Henry Dreher is the science writer who helped shape the vision for this work. His superb writing skills have brought to life the science and stories that are the heart of this endeavor. When I needed advice, encouragement, and guidance in my first attempt at writing a book, Henry supplied it with the intelligence and gentleness that are his hallmark. For this he has my great respect and sincere appreciation.

I would like to acknowledge the work of Deborah Chiel. She has worked closely with her husband, Henry Dreher, in preparing material for this book. Her expertise and dedication are greatly valued.

Janis Vallely, my peerless agent, is in large part responsible for the very existence of this book. It was she who initially called me and urged me to share my knowledge and fulfill the dream I always had. After I agreed it was Janis who smoothed out any ripples. Some of my friends advised me that writing a book could be very difficult, time consuming, and exhausting, but working with the team consisting of Steven Power, Henry Dreher, and Janis Vallely has made writing this book a most pleasant experience.

I wish to acknowledge the contribution of my sister, Catherine Haake, who read the entire manuscript and made substantive suggestions. It is my great fortune to have her and my brother, Jim, as lifelong close friends.

Dr. Yutaka Kobayashi, my former colleague at the Worcester Foundation, made significant contributions to our work. In particular, his research on blood monoamine oxidase has been crucial to the understanding of the influence of hormones on the brain.

Mrs. Mary Ellen Johnson has worked with me for thirty years. Supervising a research laboratory requires a meticulous attention to details and the intelligence and willingness to solve problems. Without her proficiency and dedication, our research would have been impossible. She is an invaluable colleague and a true friend.

I have collaborated in a number of research studies with both psychiatrists and psychologists. These include Dr. Susan Vogel, Dr. Scott Cutler, Dr. Linda G. Peterson, and Dr. Marjorie B. Snyder.

I wish to recognize my professional association with the following colleagues who have contributed to my knowledge of the effects of hormones on the brain: Dr. Mary Collins, Dr. Kenneth Gordon, Dr. Merle Ingraham, Cassandra Moribito, Dr. Deborah Sichel, Dr. Conrad Nadeau, Dr. C. B. S. Patel, Dr. Wilfrid Pilette, Dr. Paul Rosen, Dr. Tim Reiner, Dr. Pamela Staffier, and Dr. Ralph Wharton.

I wish to recognize Dr. Susan Rako and her pioneering book on the use of testosterone in menopausal women, which has accentuated an important frequently neglected portion of hormone replacement.

My early interest in endocrinology was greatly enhanced by my association with my medical school professor and friend Dr. Charles Lloyd. In Dr. Lloyd's lab I worked closely with Ms. Julia Lobotsky, who trained me in the discipline of laboratory science and became a dear friend.

Marcia Lawrence, the author of the poignant book *Menopause and Madness,* introduced me to my agent Janis Vallely. Marcia has been my strong supporter.

Mr. Peter Dellostritto has been invaluable in helping me obtain reference material for this book.

From the beginning Joëlle DelBourgo, Megan Newman, Ayesha Pande, and Nick Darrell at HarperCollins have given excellent publishing advice.

I'd like to express my thanks to Ellen Kelly-Bleakley, the artist who created figures 1, 2, 4, and 5.

Last but most important, I wish to recognize Bill and Don, to whom I have dedicated this book. My extraordinary colleague and close friend Bill Vogel and I have worked together for more than thirty years, and it is impossible for to me put into words how important our relationship has been for me. I would like to recognize his wife, Dr. Susan Raymond Vogel, his son, Douglas Peter Vogel, his daughter, Anne Hildreth Vogel-Marr, his son-in-law Jon Marr, and his grandchildren, Nathan Lee and Carey Lynn Marr. Bill has read the entire manuscript of this book and made suggestions along the way. This is what he always did for the scientific manuscripts that he, Don Broverman, and I wrote together. Don and I trusted his literary and scientific judgment implicitly. The personal-research studies referred to in this book are his and Don's as well as mine. The three of us had always been close. They were my brothers in science. Since the tragic loss of Don in 1995, Bill and I have carried on our work together and will continue to do so. More than ever I look forward to our weekly lunches together.

Don Broverman was such an integral part of my life that I feel that this book is as much his as it is mine. Therefore I would like to recognize his wife, Shirley; son, Cliff; and daughter, Sherry; son-in-law Greg Wray and grandchildren, Anya Broverman Wray and Elias Donald Broverman Wray.

My very first impression of Don Broverman was that this was a man of remarkable intellect, and that impression has remained unchanged for over thirty years. His theoretical vision of the influence of hormones on mood and cognition has now been proved to be true again and again. Like most men working together, we never put into words the mutual respect we had for each other, but it was understood. I would like to print an excerpt of a letter he wrote early one morning, sitting at his computer seven months before his death from a heart attack. I think he had just finally admitted to himself how ill he was.

"Early in life, after much thought, I concluded that the purpose of my life ought to be to contribute something meaningful to the society and civilization I lived in and benefited from. The most meaningful thing, indeed the essence of our culture and civilization, seemed to me to be knowledge. I feel that I have made significant contributions in this respect." He certainly did. This letter is one of my most treasured possessions. I miss him every day.

Hormones and the Mind

Introduction

In this book, I draw upon three decades of research and clinical experience as an endocrinologist to offer a message of hope to women: There is a neglected reason that may be contributing to the problems they are experiencing during menopause. It is a mind-healing element; it is hormones. This book has been written for women who have bouts of depression that they do not fully understand and cannot seem to resolve, even with medical treatment. Women perplexed, even stymied, by a sudden or consistent pattern of memory loss. Women who are having difficulty making simple decisions, concentrating on their work, or focusing on everyday conversations; women whose sex drive has declined, causing a strain in their relationships. If you've noticed any of these changes occurring as you approach menopause, or sometime thereafter, then *Hormones and the Mind* may provide you with clear explanations and possible solutions for your conditions. Actual case studies will introduce you to people whose lives have improved as a result of hormone therapy.

My research, and now the findings of other scientists, are pieces of a puzzle coalescing to form a clear picture: hormonal fluctuations, and particularly drops in estrogen and testosterone before and during menopause and at other junctures in the life cycle, are a major contributor to mood disorders, memory loss, mental fuzziness, and sexual problems.

Researchers in the burgeoning field of *psychoneuroendocrinology*—the interface of psychology, neurology, and endocrinology—have learned that estrogen, progesterone, testosterone, thyroid, and adrenal hormones are critical regulators of the brain. Imbalances among one group of hormones can trigger imbalances in other groups; the brain is influenced by endocrine factors that impinge on each other in a complex interplay. Without a proper balance of these hormones, we're vulnerable to mental and sexual disturbances.

How can we create such balance? *Through individualized hormone modulation.* Put simply, this means diagnosing a person's unique hormonal imbalances and correcting them with hormone preparations that restore equilibrium and therefore health—not only to the bones and the cardiovascular system but to the brain itself. *Hormones and the Mind* is a scientifically grounded approach to bolstering emotional, cognitive, and sexual health.

Most important, a series of clinical studies reveal that hormone treatment—mostly to restore natural levels—can be an extremely effective therapy for depression, memory lapses, and the loss of sex drive and function. These studies inform and reflect the successes I have overseen in the clinic. As an endocrinologist in the unique position of working in psychiatric settings, I have been privileged to help hundreds of patients overcome seemingly insoluble disorders through hormone therapy.

I have worked to bridge the gaps between medical fields long separated by false barriers of specialization. That is why I call myself a psychoneuroendocrinologist—an unwieldy title but the only appropriate one. The mind is inextricably intertwined with the nervous and endocrine systems, and I have carried out studies and practiced a form of medicine that honors these relationships. Any well-informed medical practitioner can prescribe *individualized hormone modulation. Hormones and the Mind* will describe the leading-edge research and present the stories of patients who've experienced powerful transformations as a result of hormone modulation.

Between the ages of twenty and fifty—*before* the onset of menopause—women experience a 40 to 50 percent decline in estrogen levels. At the time of menopause, when periods cease, estrogen levels drop 70 to 80 percent. The subtler changes before menopause may be enough to affect mood, memory, and sexuality. The precipitous decline during menopause surely causes psychological symptoms in women whose genes or stress levels make them susceptible. Moreover, testosterone—long known as the hormone of male sexual desire—is also the female hormone of desire. And the testosterone drop of 40 to 50 percent prior to menopause is a direct factor in the loss of sexual desire in many peri- and postmenopausal women.

These hormonal changes can cause marked shifts in cognition, mood, and behavior. Not all women experience them, but for those who do, the common experiences include feeling out of control, helpless, hopeless, irritable, forgetful, foggy, sexually inert, or some tormenting combination thereof. Hormone modulation is an answer for many of these women, who often require multiple-level adjustments of their endocrine profile.

One of the most exciting new developments is the finding that estrogen, testosterone, and thyroid hormones can work in a complementary or even synergistic fashion with antidepressants, including Prozac and its chemical cousins. In many cases, these hormones, especially estrogen, enable antidepressants to work for patients who've gotten variable or no relief from these medications alone. *Hormone modulation may provide a solution for legions of depressed women on the antidepressant merry-go-round.*

Any woman with mood disorders, severe PMS, forgetfulness, confusion, difficulty making decisions, concerns about Alzheimer's, or waning sexual desire or capacity should investigate her hormonal profile. Women and their partners should also be aware that any man with hard-to-treat depression and sexual dysfunction may benefit from hormone treatments.

THE ORIGINS OF HORMONE MODULATION

My own scientific journey toward hormone modulation for the mind began with an unusual observation—that male research subjects with contrasting cognitive abilities had differing hormone levels.[1] Specifically we discovered that our male subjects with more body hair tended to have higher testosterone levels and they also performed better on certain cognitive tests. The relationships between hormone levels and cognitive abilities in these men provided me with an early clue to the effects of hormones on the brain. Spurred by this finding, and by an intriguing though neglected body of evidence dating back decades, I set out to understand the influence of hormones on the brain and to exploit that influence for the well-being of my patients.

My lab research on hormones and the brain began in the late 1960s and early 1970s, while I was Chief of Clinical Research in the Department of Biological Research at Worcester State Hospital in Massachusetts. My colleagues, Dr. William Vogel and the late Dr. Donald Broverman, and I demonstrated for the first time that the loss of estrogen causes a rise in an enzyme, monoamine oxidase (MAO), that has crucial effects on the brain and on mood states.[2] We later showed that estrogen treatment normalized

MAO—and restored emotional well-being to women plagued by depression. During the late 1970s, we conducted a double-blind clinical study at Worcester State Hospital that was sponsored by the National Institute of Mental Health (NIMH), treating women with severe depression—inpatients who had not responded to any form of treatment—with high doses of estrogen. We showed that estrogen reduced depressive symptoms and resulted in full-scale remissions in many so-called hopeless cases. When I witnessed women who'd been hospitalized for years finally leave the institution, I felt strongly that my colleagues and I had come upon a significant medical development, one that would eventually change the lives of innumerable women for the better.

In the years hence, as Senior Scientist at the Worcester Foundation for Biomedical Research, and currently as president of the Broverman Research Foundation, I've continued to study and apply hormone modulation for mood, memory, and sexual health. I have found that hormone modulation can restore natural balance to the endocrine profiles—and the emotional lives—of suffering women. Hormonal *rebalancing* restores the mental capacities and emotional equilibrium we need to lead joyful, fulfilling lives. As such, this form of treatment respects the natural biological order of mind and body.

Hormone rebalancing is not reserved for women. Hormones affect the brains of women and men, though in different ways. As such, hormone modulation can be used to treat mood, cognition, and sexuality in both genders. While this book focuses on women, another whole book could easily be devoted to hormones for men, whose emotional states, mental functions, and sexuality are just as influenced by hormones. In this book, with its focus on women, I devote a section in chapter 4 on sexuality to the ways that women understand and help male partners or friends with hormone-associated problems.

While estrogen is a key, individualized hormone modulation for women may also include testosterone for emotional and sexual health, and thyroid replacement for patients with imbalances that cause cognitive or emotional symptoms. Surprising numbers of menopausal women with emotional symptoms have hypothyroidism—low thyroid levels. I, therefore, find it necessary to treat 10 to 15 percent of my patients with the thyroid hormone called thyroxine.

Hormones and the Mind offers an individualized plan on how to enhance mood, memory, mental sharpness, and sexual vitality with hormone therapy. I offer recommendations and detailed case histories on the use of hormones to treat or prevent:

- *Depression and anxiety*
- *Mood disorders associated with perimenopause, menopause, PMS, and the postpartum period*
- *Cognitive impairments, including mental "fuzziness," difficulty concentrating and making decisions, and verbal slips*
- *Short- and long-term memory lapses*
- *Alzheimer's disease and dementia*
- *Sexual disinterest and dysfunction*

In chapter 1, "Deep Impact," you will learn about the newest scientific evidence for a paradigm-shifting concept that hormones have a major influence on the brain. Throughout the chapters on mood (2), cognition and memory (3), sexuality (4), PMS (5), and Alzheimer's disease (6), you will read about clinical studies and cases showing that hormone therapy can restore mental and sexual health. (Consult the References section for scientific citations.) In chapter 9, "The Safety Equation," you will find clear advice on how to balance the risks and benefits of hormone modulation, enabling you to optimize your health with minimal risks. And in chapter 10, "Taking Charge of Your Hormones," you will learn how to arm yourself with knowledge, so that you can inform and mobilize your doctors and other health care providers in order to develop a program of hormone treatment that is best for you.

HORMONE THERAPY IS NOT "UNNATURAL"

Some women are concerned about taking hormones, based on the view that if "nature" meant our hormone levels to decline at midlife, then we should not tamper with this process. It may indeed be natural for estrogen and testosterone to decline at midlife, but is it desirable? And if everything natural is good for our health and well-being, why then do heart disease rates increase dramatically when women's estrogen levels drop? Surely, that's not good for women's health and well-being.

In the case of menopause-associated suffering, we may be dealing with the natural result of an unnatural reality—the longer lives we live as we enter the twenty-first century. In the previous century alone, medical technology had added decades to the life expectancy of women.

When women did not live far beyond their fifties or sixties, there was less concern about the symptoms of menopause. (Not only that, emotional distress, mental lapses, and sexual problems were not openly discussed in those days.) Women's bodies may not have been "designed" to live for a lengthy period of time after the ovaries shut down.

That being the case, the medical experts who've made such longevity possible have a responsibility to make certain that these extra decades of life are of good quality. Ovarian decline is certainly natural—it's not a disease. Moreover, some women have no serious disturbances of mood, cognition, or sexuality that require treatment. But many women do suffer with mood disorders, cognitive impairment, sexual dysfunction, distressing physical symptoms, osteoporosis, or incipient heart disease. While menopause is a natural process there is nothing intrinsically healthy about the suffering some women experience as their estrogen levels decline. These women need and deserve medical treatment to ease these ailments. The question is: What treatment is best?

As I discuss in chapter 7 and elsewhere, I certainly favor any conventional or alternative therapy proven to ease the symptoms and prevent the dangers associated with hormonal flux at midlife. Unfortunately, in many cases there is too little evidence that such treatments are as reliably effective as hormone therapy. Also, individualized hormone modulation is a form of treatment that utilizes natural substances—hormones—in order to restore the natural endocrine balance that women enjoy before menopause.

Unlike standard, "one size fits all" hormone-replacement therapy (HRT), individualized hormone modulation uses different preparations and dosages of hormones, and continual monitoring of blood hormone levels, to enable women to experience the health benefits of a normal hormone profile. We now know that those benefits include mental, emotional, and sexual health—in addition to bone health, breast health, and heart health. Individualized hormone therapy does not push women's hormones to super-high levels, a truly unnatural manipulation that would undoubtedly cause adverse effects and long-term health risks.

THE TRUTH ABOUT CANCER RISKS

Many women considering estrogen-based treatments are concerned about breast cancer risks. In this book, I explain why the media has overstated these risks, and I try to put them into a proper, balanced perspective. Every woman must make her own decision based on her unique personal and medical history, and I believe she must consider any risks alongside the benefits of treatment. But she should also know the real facts about breast cancer and other risks, as we now understand them. Few women know, for example, that:

- The overall evidence on the breast cancer risks associated with estrogen is equivocal: Some studies suggest a small increased risk after many years,[3] while other studies show no increased risk whatsoever.[4]
- The largest recent study of 60,000 nurses demonstrated a 24 percent reduction in breast cancer mortality among estrogen users during their first decade of use.[5] The risk went up only after a full decade on treatment. Overall, the chances of dying from any cause *went down by 37 percent* during the first decade, and it continued to be 20 percent lower after ten years.
- A 1999 published study of more than 37,000 women showed no linkage between estrogen use and the potentially lethal forms of breast cancer that account for 85 to 90 percent of all cases. There was a small increased risk of less common cancers that are much more easily treated and cured.[6]
- Contrary to popular opinion, only 3.8 percent of American women who live to ninety will get breast cancer, while almost 50 percent will die of heart disease. Estrogen cuts the risk of developing heart disease by approximately 40 to 50 percent.[7]

Estrogen-based hormone modulation not only cuts heart disease risk nearly in half, it reduces the risks of osteoporosis,[8] colon cancer,[9] stroke,[10] macular degeneration,[11] and Alzheimer's disease[12]—not to mention the reduction in mood,[13] cognitive,[14] and sexual disorders[15] that is the focus of this book. Here, you will also learn how you can take forms of estrogen along with progesterone, and in some instances, other hormones, that involve the least risk of short- and long-term adverse effects.

THE PROMISE OF HORMONE MODULATION

Women and their doctors should never minimize the distressing and sometimes severe emotional and cognitive symptoms associated with hormonal flux. I encourage all women to take their own symptoms seriously, conduct their own investigations, and to be dogged in their search for medical guidance and effective treatment. The quality of their work lives, family lives, and emotional well-being may be at stake.

Some women worry about the investigation or even discussion of mood, memory, and cognitive disorders associated with the menopausal transition. They fear the stereotypic notions that women dealing with hormonal flux might be less functional, competent, or reliable on the job. But we should have long ago shed these shopworn prejudices. It is time to let go of our trepidation about addressing a real set of medical problems women face, so that we can solve them. A stellar example of someone who did just that is Patricia Ireland, head of the National Organization for Women, who admitted having cognitive lapses during her own menopausal transition. She would occasionally lose the ability to find the right word and forget appointments. But these lapses cleared soon after she began hormone therapy. Now, Ireland is fully in favor of research and public information on this issue. "All of us who are honest in our philosophy and our politics have to be willing to look at what the studies are showing," she said.

Many women in the twenty-first century face problems linked directly to longer lives in the modern medical age. In this book, I offer strong evidence and my own thirty years of clinical experience to support this new approach, one that regards women not just as a collection of body parts, but as whole human beings with minds and bodies that interact. When women are able to get treatment for their whole selves, they can enter this new era expecting not only a longer life span but a more productive, joyous, and richer life span. That is the promise of individualized hormone modulation.

1

Deep Impact

HORMONES AND THE BRAIN

PAM'S STORY

Pam woke with a start. Her heart was pounding and she had the horrible feeling that something was seriously wrong. She put her hand to her chest to feel her heart beating. At first, the beat was just rapid, but then the pounding became irregular and her breath came in short gasps. She thought, "My God, I'm having a heart attack. I'll wake Jack and have him call an ambulance." Her mind was racing with panic. Pam tried to rein in her thoughts, but there was no stopping this train of fear and loss of control. The panic became so intense that her body was covered in perspiration, soaking her nightgown. "I'm going to faint," she thought.

Here was this forty-four-year-old, high-powered executive, president of her own company, sitting in my office telling me this story of her panic attack three nights earlier. Pam was experiencing symptoms she could not understand—not just the panic, but also severe moodiness, irritability, and forgetfulness that were wreaking havoc in her marriage and work life. She wept at the slightest provocation. "At one point I was watching people singing Christmas carols on television," recalled Pam. "When they sang 'O Little Town of Bethlehem,' I dissolved into tears. It was absurd. There was something very wrong and it was affecting not only my body but my mind."

Pam wondered whether there was a link between her symptoms and menopause, which is why she came to see me. Her symptoms followed a pattern familiar to me, but I needed to hear more of her story.

In the midst of her panic she had gotten out of bed, walked into the bathroom, and closed the door so the light wouldn't wake her husband. The face in the mirror wasn't familiar. It was pale and frightened. Her eyes looked as if she were about to cry. Her anxiety surged, and the panic was once again uncontrollable. She raised her fist and lunged forward, stopping just short of crashing her hand into the mirror. Pam thought she might wake Jack, hoping that he'd hold her and make the fear go away. He'd always calmed her down in the worst crises, at least in the past. Jack had been her refuge.

Lately that closeness had changed. Pam could get angry with Jack in a flash, and for seemingly no important reason. Of course, this would cause him to withdraw. Tolerance was never one of Pam's outstanding traits, but now, when the children would lose things, she'd fly off the handle. What was happening to them? What was happening to her? Last week, she couldn't find her car keys for three whole days. She finally found them when, in search of a missing earring, she opened the top drawer of her bedroom dresser. It was where she'd always put things whenever she was in her cleaning mode. "Normally," she said, "I'd have retraced my steps and found them right away."

The memory loss really frightened Pam. Her mother had Alzheimer's disease, and she was terrified at the mere thought that she, too, could be afflicted. She remembered how maddening it had been to answer the same question over and over again, because her mother didn't remember what Pam had said moments earlier. Pam worried that her memory lapses might be the beginning of a much larger problem. "Lately at work, when my secretary asks for a decision on a simple problem I can't make up my mind. That's not at all like me. This whole thing is a nightmare. I feel like I'm losing my mind."

I've heard that desperate statement—"I feel like I'm losing my mind"—repeatedly from women who are passing through menopause. Based on my years of research and clinical practice, I can reassure them that if their problem is a hormonal imbalance, the problem can be rectified, and once it is, their many painful and disruptive symptoms will abate. Pam had offered many clues that her difficulties could be attributed to hormonal imbalance—namely, the combination of night sweats, insomnia, panic attacks, weepiness, and memory problems. These are all classical symptoms of estrogen deficiency associated with menopause.

But there was one area Pam hadn't touched upon. I asked her, "How have the sexual relations between you and your husband been going?" She paused before answering, "I'm glad you asked. I was a little reluctant to raise the subject, but the other night Jack and I talked about this. He and I have been avoiding each other lately, and we haven't made love in God knows how long. I told him I had lost interest in sex, and the last time we made love it was painful. He remembered that it had been uncomfortable for him as well, and he was worried that some of my sexual disinterest might be his fault. Both of us kept saying, 'It's not you, it's me.' "

"This is not your fault, nor is it your husband's," I said. "Hormones affect the brain. Your symptoms may result from a hormone imbalance, and hormone imbalances can be treated." This is what I tell my patients, and it is often the first glimmer of hope they've had after a long period of suffering.

Pam had come to me for a consultation because she suspected that her symptoms—including her cognitive, mood, and sexual difficulties—could be caused by hormone fluctuations, perhaps from early menopause. Pam and Jack had long assumed she was too young to be menopausal, but they felt they had to find out if there was a biological cause and solution to their mutual misery. Jack summed it up perfectly during their late-night talk: "We are both getting older, but it's too early for this."

During that first consult, I confirmed for Pam that her symptoms were likely related to the menopausal loss of two ovarian hormones, estrogen and testosterone. But to properly diagnose and treat her, I had to test her hormone levels. Sure enough, the results confirmed her low estrogen and testosterone levels were signs of perimenopause, that period of hormonal fluctuation prior to menopause. I started her on estrogen-replacement treatment, and within one week, her anxiety and depression began to subside. Within six weeks, Pam's decision-making ability, short-term memory, and general good spirits had completely returned. Her personal physician was astonished by her turnaround.

Oral micronized progesterone capsules were also added to Pam's therapy for ten days each month. Progesterone is the ovarian hormone necessary to protect the uterus against cancer when estrogen is being administered. I also prescribed a low dose of testosterone for her loss of sexual desire. The results were immensely heartening for Pam and Jack. "I could not believe that a little bottle of pills had done this for me," she said. "Three weeks ago I didn't care if I had sex at all, and when we did, it was such an effort to climax it was hardly worth it. Now, I'm excited about making love and afterward I feel fulfilled. I can't remember how long it's been since I felt this kind of sexual satisfaction."

Pam's story is one of hundreds, all of them told to me by patients who have come to see me for a complex of symptoms involving mood, thinking, sexual functioning, or unexplained physical disorders. We are all familiar with the hot flashes and vaginal dryness associated with menopause. But cases like Pam's reveal that hormone imbalances, including estrogen and testosterone deficiency, are behind a much more encompassing set of problems faced by women, not only after menopause but during the menopausal transition. This oft-misunderstood "perimenopause" is now recognized as a period lasting years, in some cases starting in the early forties until a woman's last period in her late forties or early fifties. Depression, short-term memory loss, and loss of sexual desire or function are all too common during this transitional time, and afterward.

The field of research devoted to the study of hormones on the brain is exploding, because we now understand that hormones play a fundamental role in neurotransmission—the interaction of brain chemicals with the nerve cells of the brain.[1] This is crucial, since neurotransmission is the scientific name for the brain processes that determine how we think, feel, and behave, and how we function as sexual beings. Estrogen, progesterone, testosterone, and the thyroid hormones are as essential to our moods and cognitive abilities as food-based nutrients are to our basic cellular functions.

While the role of hormones in the brain has been studied for decades, the clinical use of hormones to treat patients with mental illness has not kept pace with the knowledge we have gleaned from research studies involving both humans and animals. In spite of the growing body of evidence that hormones affect mood, learning, and sexual function, the general medical field has almost totally ignored hormonal approaches as treatments for these disorders.

Today, depression is primarily treated with the current popular antidepressants including Prozac, Zoloft, and Paxil. These SSRIs (Selective Serotonin Reuptake Inhibitors) can be extremely effective, but a subset of people—often, women over sixty—do not respond to them.[2] Clinically I have found that many women in their forties and fifties have this problem as well. In many cases, the SSRIs effectively lift depression, but the person may lose sexual desire or function. Other antidepressants used to treat depression and anxiety, including monoamine oxidase (MAO) inhibitors and tricyclic antidepressants, can also have adverse effects on sexual function, often by delaying orgasm.[3] Largely because of the problems with conventional antidepressants, there has been increasing interest in so-called natural treatments for depression, including St.-John's-wort, 5-HTP, SAM-e, and ginkgo biloba. B vitamins, including B_6 (pyridoxine) have been touted as nutritional supplements that may relieve depression.

Meditation, yoga, and acupuncture are other nontraditional approaches that have also been used to treat mental and emotional disorders. While there is some limited scientific evidence to support these natural treatments, more data are required for them to pass muster. Yet the media have touted natural therapies in what amounts to a revolving remedy-of-the-month. Even the best-tested chemical antidepressants fall short, in the sense that a definite subgroup fails to respond, and some responders eventually relapse. For most SSRIs, approximately 33 percent are nonresponders after eight weeks of treatment. However this figure may vary depending on the population being studied and the degree of response being evaluated.[4] Some other factors, heretofore ignored, may be the basis of treatment failure of disorders of mood, memory, and sexual function. In my view, hormones represent this "missing link."

There are new classes of drugs to treat cognitive impairments (mainly poor concentration and memory problems), such as Cognex, and medications for sexual dysfunctions, such as the wildly popular Viagra. While these are promising medications, and Viagra is obviously a success, mainstream medicine still hasn't grappled with the causes of cognitive and sexual dysfunction, particularly among peri- and postmenopausal women after the age of forty. Nor have they explored the vast potential of hormone treatments for women who are struggling with these serious, quality-of-life hampering problems.

So ultimately, despite the great advances in psychopharmacology and psychiatry, many people—especially women—are not getting the help they need for their cognitive, emotional, and sexual difficulties. Moreover, certain hard-to-treat conditions—including Alzheimer's disease[5] and Parkinson's[6]—appear to have a hormonal component and a set of potentially effective treatments that have also been largely ignored. We need to thoroughly reassess the significance of hormones in all of these conditions.

DEEP INTO THE BRAIN

Before I explain how and why hormone modulation is the solution for many people with these symptoms, I'll take you on a journey deep into the brain, where hormones are essential players in the fundamental communication pattern of billions of nerve cells. At first this journey may seem rather technical, but persist and your effort will pay off, giving you a better understanding of the inner workings of the brain.

. . .

In their roles as regulators and stimulators, hormones are responsible for brain cell functions that determine mood states, memory storage, cognitive capacities, and sexual drive. There is a long, albeit spotty, history of interest in hormones and the mind. The medical literature of the late 1930s includes references to mood disorder as a symptom of the menopausal loss of estrogen, and a few reports of estrogen being used to treat depression. One of the diagnostic terms bandied about was "menopausal melancholia," an apt if not terribly illuminating or precise description. But those early researchers abandoned the effort because they had imprecise ways of measuring estrogen levels in the body. They would treat patients with estrogen and look for changes in vaginal cells to find out whether estrogen levels had increased. With low doses of estrogen, the vaginal cells changed ("cornified") quite readily. When researchers did not see marked behavioral improvements in their patients, they assumed that estrogen was getting into the system, but it simply wasn't working. What they didn't know was that vaginal cells have much higher affinity for estrogen than brain cells, so these doses probably had minimal impact on the brain, which would explain why there was little benefit for depressed patients. In those days, endocrinologists couldn't easily test for estrogen blood levels—a much more accurate way to determine how much of the hormone is getting into the system, and hence, the brain.

The effort was largely abandoned, and it became even more suspect when scientists questioned whether hormones could even get *into* the brain. When I first became interested in this area, in the mid-1960s, most biomedical scientists were convinced that hormones could not pass through the seemingly impenetrable portal known as the blood-brain barrier. I recall an editorial in one of the leading medical journals that said this barrier, which presumably blocked any hormones from entrance, made this whole field of investigation a waste of time. However, by the late 1960s, several investigators, most notably Dr. Donald Pfaff of Rockefeller University, tagged hormones with radioactive isotopes, injected them into laboratory animals, and used imaging technologies to scan their brains.[7] The researchers were astonished when the scans revealed brain structures that clearly glowed—luminescent proof that hormones had infiltrated brain tissues. That was just the beginning of a long, substantive, barrier-breaking scientific investigation of hormones and the mind.

My own research into the effect of hormones on the brain began in the late 1960s and early 1970s at the Worcester State Hospital where I was fortunate to meet and begin a lifelong collaboration with my two colleagues, psychologists William Vogel, Ph.D., and Donald Broverman, Ph.D. We conducted a series of early studies that led us down the challenging, often circuitous, and sometimes thrilling path of scientific discovery.

At the time, I was invited to a meeting in a conference room at the Worcester State Hospital, where I first met Drs. Vogel and Broverman. They were conducting studies on the cognitive styles of men, a subject I did not know much about. I had the rather typical attitude of a medical doctor toward psychologists—one of mild condescension. I assumed that their scientific acumen would be deficient, that their measures and ideas would lack the precision of hard biological science.

I was dead wrong, but at least I was open enough to realize it pretty quickly. Vogel and Broverman were superb scientists, as good or better at research design and statistical analysis than most M.D.s. Moreover, they were interested in bridging psychology and biology, and I soon realized that I, too, was fascinated by the linkages between mental and biological functioning. As a research and clinical endocrinologist, the biological realm that obviously captivated me was hormones, and so I was receptive to the idea of a bridge between hormones and the mind. While our early work involved testosterone and men, I hypothesized from the start that "gonadal" hormones produced by our sex glands—whether testosterone or estrogen—might influence the brain in both men and women. Our initial studies on men were to *prove the principle,* which would enable us to move to research on estrogen and women.

How Hormones Influence the Brain: The Molecular Drama of Mood and Cognition

Neurotransmission is the process by which chemical impulses are transmitted from one brain cell—neuron—to another. This neuron-to-neuron communication is the foundation for everything that goes on in the brain, the biochemical essence of all our thoughts, behaviors, and emotions.

The brain contains billions of neurons, which are composed of a cell body (like any other cell, with a surface membrane and a nucleus); a set of fibers, known as dendrites, that branch out from the body, making up the dendritic tree; and one long, thin, fingerlike tube, called the axon, that extends out from the cell body. (See Figure 1.) Nerve cells communicate in

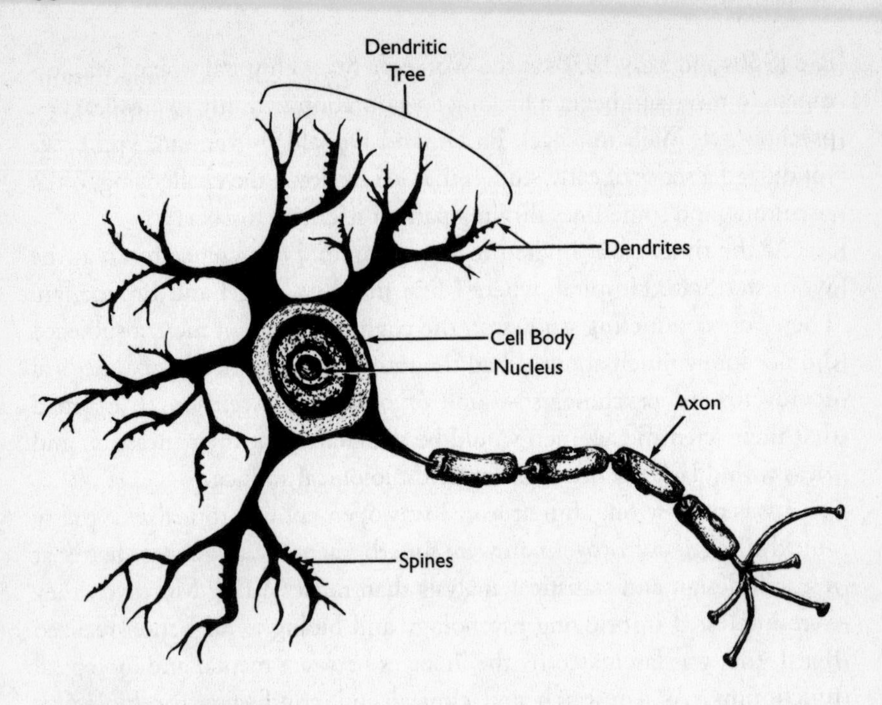

FIGURE 1 Neuron
The neuron has a cell body, an axon, and a dendritic tree that consists of branching dendrites. This tree is constantly changing and revising its connections with other dendrites throughout life.

various ways, including through electrical impulses sent and received by the dendrites. Most important, nerve impulses traverse the cell body down the axon, which appears to connect at its end with the axon from another neuron. In fact, the end of the axon does not actually touch the end of another axon; there is a small separating space called the synapse.[8]

Thus, what occurs in the synapse is critical, but there is a lot scientists don't know about neurotransmission. For instance, we suspect that there are upward of a hundred or more different types of neurotransmitters but have yet to characterize most of them. However, neuroscientists and psychiatrists have identified neurotransmitters that appear to be critical to the workings of the mind, including norepinephrine, serotonin, dopamine, acetylcholine, and GABA. And they have determined, with some specificity, how neurotransmitters act within the synapse.

The neurotransmitter molecule travels across the synapse from the end of one neuron to the start of another. After it crosses the synapse it is picked up on the other side by a "receptor" protein that sits on or underneath the surface of the neuron. (See Figure 2.)[9] The configuration of the neurotransmitter is like a key that fits into a corresponding lock, the receptor. Once the lock-in-key connection occurs, the chemical "information" is passed on to the neighboring neuron. When neurotransmitters lock into their receptors, the "message"—whatever it may be—is passed along from neuron to neuron, and the business of the brain is conducted as it should be.

If neurotransmission is insufficient or interrupted, communication breaks down, or is less than optimal. Communication between neurons makes possible communication between parts of the brain—such as the structures that control thinking and those that control speech. Optimal neurotransmission is therefore the basis for healthy mental functioning on every level.

Neurotransmitter messages can be weak or strong, and this depends partly on whether the molecules remain in the synapse where they bind repeatedly to their corresponding receptors. Other factors are critical, too, such as whether there is enough neurotransmitter present, whether the molecules bind tightly to the receptors, and whether there are enough receptors present to send strong signals.

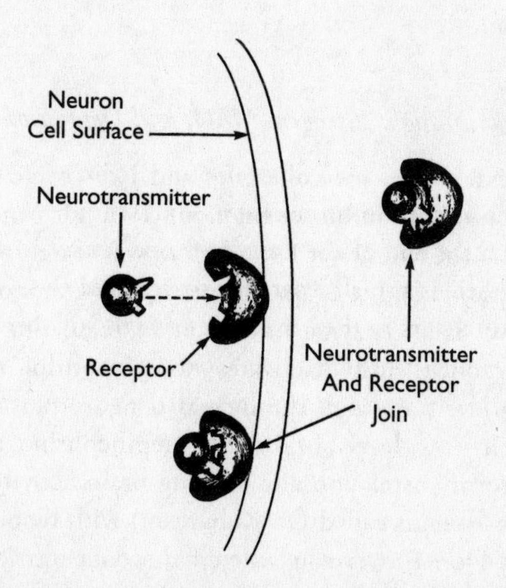

FIGURE 2 Neurotransmitter-Receptor Interaction
The receptors ("the locks") are where the neurotransmitters ("keys") bind to the cells. Some binding sites are on the surface of the neuron (*left*) and others are beneath the surface (*right*).

One of the most significant developments in neuroscience and psychiatry in recent decades has been the recognition that neurotransmitters play key roles in the drama of our moods and emotional states. Norepinephrine and serotonin are two major neurotransmitters that influence mood.[10] If there is insufficient or inadequate binding of these two neurotransmitters to their intended receptors on neurons, we may be prone to depression, manic-depression, and other mood disorders.[11] Even the tendency toward violence has been associated with impairments in the transmission of serotonin. Deficiencies in these neurotransmitter systems may also contribute to cognitive impairments.

The most commonly prescribed antidepressants, such as Prozac, Zoloft, Paxil, are known as the Selective Serotonin Reuptake Inhibitors—the SSRIs. These medications specifically block the nerve cell from taking back serotonin—preventing so-called "reuptake." As a result, more serotonin molecules "remain" in the synapse, where they're more likely to be "received" by the neighboring nerve cell. For someone who is clinically depressed, anything that allows more serotonin to remain in the juncture between neurons is a biochemical blessing.

Mood, memory, and other cognitive ability depend upon proper neurotransmission. We will next examine the role of hormones and neurotransmission.

An Early Breakthrough: Estrogen, MAO, and Depression

It was an accident that my colleagues and I discovered how estrogen and testosterone might influence brain function. Dr. Yutaka Kobayashi, a biochemist at the Worcester Foundation, was searching to determine if the blood contained a brain chemical called monoamine oxidase (MAO). MAO is an enzyme important in regulating brain activity. MAO breaks down and inactivates norepinephrine and serotonin. When MAO levels are high the increased neurotransmitter inactivation results in low levels of brain norepinephrine and serotonin, impairing neurotransmission and slowing brain activity. We supplied "Koby" (as his friends called Dr. Kobayashi) with blood samples from our patients. Don Broverman entered the data into a vintage computer at MIT, and we discovered very strong statistical relationships between low levels of MAO, and high levels of estrogen and testosterone.

What did this mean? We surmised that the two hormones inhibited the action of the MAO enzyme. This would *bolster* neurotransmission involving norepinephrine and serotonin—explaining why these men were better at automatic tasks in our laboratory studies.

This information became critical when we learned that depressed women had high MAO levels in their blood. When we gave estrogen to the depressed women the MAO levels declined and their moods improved. We received support from the National Institutes of Health (NIH) to conduct a trial of estrogen treatment on forty severely depressed women. These women had failed to respond to the usual antidepressants. A number of these women underwent a complete remission from their intractable mental illness. I will discuss this study in detail in chapter 2.

Estrogen's Effects on the Brain

The following are a list of the actions that estrogen has on brain function in addition to its effect on MAO:

- Estrogen increases the amount of mood-regulating neurotransmitters available in the brain—including norepinephrine, serotonin, and acetylcholine.[12]
- Estrogen increases the density of neurotransmitter receptors on the surface of nerve cells, ensuring the "key-in-lock" fit that is needed for proper neurotransmission and mental well-being.[13]
- Estrogen helps maintain the integrity of dendrites, the delicate branching structures of the brain that conduct electrical impulses and release chemical neurotransmitters, allowing neurons to communicate with one another. The hormone keeps dendritic spines strong and well defined.[14]
- Estrogen acts in concert with growth factors—which enhance the growth and survival of nerves within the brain. Estrogen deficiency may contribute to the atrophy or death of these nerve cells. This has implications for the loss of brain tissues associated with aging and Alzheimer's disease.[15]
- Estrogen increases the ability of neurons to "connect" in various parts of the brain, regulating emotional states such as depression and anxiety.

Neurons, Blood Flow, and the Hormone-Brain Connection

I've emphasized the effect of hormones on neurotransmitters and their receptors, but there's much more to the story of the deep impact of hormones in the brain. Neurons themselves appear to depend on estrogen for their structural integrity. Figure 3 presents starkly illustrative photographs of a culture of nerve cells before and after they were bathed with estrogen. The photos show that the presence of this hormone helps to maintain the strength and integrity of our brain cells.

Estrogen is also necessary for proper blood flow to parts of the brain that are responsible for emotion, memory, and cognitive function. The research on the role of hormones in the blood flow to the brain is not being conducted on the outer fringes of neuroscience. A number of leading scientists at the government's National Institute of Mental Health (NIMH) have demonstrated that estrogen increases blood flow to the brain. A sophisticated technique (proton emission tomography—PET) was used to evaluate regional blood flow to the brains of women who performed a mentally challenging task. They discovered that dur-

Control **Estrogen**

FIGURE 3 Neuronal Response to Estrogen
A clear example of the effect of estrogen on the brain. Two embryonic nerve fibers from the brain of a rat grown without (*left*) and with (*right*) estrogen exposure show many new neuronal connections after estrogen exposure. (*Photos by Dr. D. Toran-Allerand. Reprinted by permission of* Brain Research.)

ing this task, if estrogen levels were low, there was a striking decrease in blood flow to the subjects' brains. When the women were given estrogen or progesterone, the PET scan showed a sudden dramatic normalization of blood flow in the cerebral cortex. The PET scans of women taking hormones show bright patches signifying substantial blood flow to the cerebral cortex during the cognitive task.[16]

At Yale University Dr. Sally Shaywitz and her colleagues conducted a randomized trial to find out whether estrogen treatments influenced the patterns of brain activation in forty-six postmenopausal women as they carried out a series of verbal and nonverbal memory tasks. When the women were receiving estrogen (1.25 mg of conjugated equine estrogens—Premarin), imaging technology demonstrated that parts of the brain responsible for higher mental functions became highly active. These included a portion of the cerebral cortex. Their study was published in 1999 in the prestigious *Journal of the American Medical Association,* and in their conclusions, Shaywitz and her associates wrote:

> Estrogen in therapeutic dosage alters brain activation patterns in postmenopausal women in specific brain regions during the performance of the sorts of memory function that is called upon frequently during any given day. These results suggest that estrogen affects brain organization for memory in postmenopausal women.[17]

Alzheimer's disease is associated with a reduced cerebral blood flow. Since estrogen has been known to increase cerebral blood flow, this may be one of the mechanisms by which estrogen can prevent the development of this disease.

Estrogen and Its Receptors: Keys to Mood and Cognition

Consider the fact that all of estrogen's ubiquitous mood-modifying and mental-sharpening abilities depend on both the hormone itself and its receptors throughout the brain. It's not hard to understand then how chronically low levels of estrogen will cause emotional and mental difficulties in at least a significant subset of women. Prior to the actual onset of menopause, estrogen and testosterone levels, gradually over three decades, decline 40 to 50 percent. At the time of menopause estrogen levels plummet another 70 to 75 percent. This down-rushing elevator can complete its crash in a matter of months, sometimes wreaking psychological and physical havoc. We know that estrogen replacement can offset the physical symptoms, the hot flashes and vaginal dryness. But we now know from a combined analysis of twenty-six studies that proper estrogen treatment can also effectively reverse depressed mood in these women, and that the addition of testosterone makes treatment even more effective.[18]

Dr. Barbara Sherwin, a professor of psychology at McGill University in Montreal, has carried out numerous studies demonstrating the marked benefits of estrogen, and combinations of estrogen and testosterone, on the mood states of women with both surgical and natural menopause.[19] Dr. Sherwin has also shown that estrogen facilitates short-term memory in menopausal women.[20] Other researchers, including myself, have delved into the effects of estrogen and testosterone on cognitive functioning. Our most recent study, currently being prepared for publication, demonstrates that estrogen treatment allows menopausal women to make decisions more accurately and with greater confidence than when they are treated with a placebo. That is why estrogen treatment helped Pam's decision-making ability. Throughout this book, you will learn about a vast number of studies on the positive effects of hormonal treatments on mood, memory, sexual function, and a surprising range of diseases—research that is setting the stage for the new role that hormones may play in the practice of medicine and psychiatry.

LAURA'S STRUGGLE WITH TREATMENT-RESISTANT DEPRESSION

As I have emphasized, estrogen not only boosts neurotransmitters, it also stimulates their corresponding receptors, the molecular keyholes essential for the cellular connections we need for proper function. Unless there are enough serotonin-specific receptors on the neurons, all the serotonin in the world won't make a difference to a woman's mood state. Without estrogen, these brain cell receptors literally atrophy. Therefore, Prozac given to a woman with low levels will not have an antidepressant effect, even though it increases serotonin levels. It is as if we have all the keys necessary to open the multiple locks of an apartment building only to find that the keyholes are missing. Now, there is a great deal being written and said lately about the failure of SSRIs to work in a substantial subset of patients. It is my firm belief that many women either don't respond to these medications, or they relapse after a period of time, *because low estrogen diminishes the number and quality of serotonin receptors in the brain.* The evidence to support this statement is beginning to amass, and I'll provide it in the next chapter. It was this phenomenon of missing keyholes that probably prevented one of my patients, Laura, from responding to treatment with antidepressants. Laura's story demonstrates that combined estrogen and testosterone can induce an antidepressant effect in a menopausal woman previously unresponsive to antidepressants.

The first time I saw Laura, she told me that her last period had occurred two years earlier. She had occasional hot flashes, and night sweats with insomnia, but her primary symptom had been depression. To her, getting out of bed in the morning was like lifting a gargantuan barbell. The doctor who referred her was a wonderful psychiatrist from Boston, Dr. Mary Collins; we'd become friends after jointly treating a number of depressed menopausal patients. Dr. Collins had been treating Laura, who had a family history of serious depression, for the past four years. Her mother and two of her brothers had been diagnosed as manic-depressive, and her father was a suicidal alcoholic who required antidepressive treatment.

Initially, Laura had been given Prozac in gradually increasing dosages. But after three or four months, it seemed to stop working, and she was put on another antidepressant, Wellbutrin. After six months when the Wellbutrin was clearly ineffective, Dr. Collins referred Laura for a hormonal evaluation. At the time of her initial visit, her blood estrogen levels were extremely low, and so were her testosterone levels.

She explained tearfully that in the past eighteen months her mood had been worsening in spite of her medication. Laura was not only depressed, however; she was having extreme difficulty concentrating at work. Moreover, Laura's relationship with her boyfriend was deteriorating, because she had no real interest in sex. Indeed, she was unable to achieve orgasm. I was quite certain that her sexual symptoms were caused by low blood testosterone levels. The small amounts of testosterone in women's bloodstream and brain have a major stimulating effect on their sex drive, and I believed that her low blood testosterone contributed to her sexual dysfunction.[21]

I considered the likelihood that Laura's low estrogen and testosterone levels, characteristic of menopause, were the underlying reason she failed to respond to her antidepressants. Again, through one mechanism or another, most antidepressants work by increasing neurotransmitters in the synapse, and these brain chemicals can't do their jobs unless neurons have the necessary quantity and quality of receptors. Laura's estrogen deficiency probably caused atrophy of her receptors—especially those for serotonin—thus explaining her poor response to Prozac.

I prescribed estrogen and testosterone for Laura, while she remained on antidepressants. Over a period of several months her depression lifted and her concentration improved, decreasing her stress at work. After six months of treatment, Laura was sleeping seven to eight hours a night. Her depression was markedly relieved, and she had regained her powers of concentration. Instead of being overwhelmed by the task of getting out of bed, Laura awoke each morning with energy and enthusiasm. Her outlook changed so significantly that she took on the challenge of a new job. While we don't yet have the technology to readily prove this, I am quite certain that estrogen treatment helped to restore the number and density of Laura's neurotransmitter receptors, thus enabling her to respond—quite dramatically—to antidepressant medication.

Hormonal modulation with both estrogen *and* testosterone was necessary to relieve Laura's other symptoms. Her body no longer was tormented with hot flashes and soaking night sweats no longer awakened her. The insomnia that had disrupted her sleep decreased and gradually disappeared. At the same time, Laura experienced a renewed interest in sex, and she was once again able to experience a normal orgasm. The concurrent improvements in Laura's sex drive and the improvement in her mood were not coincidental, but reflected the combined positive effects of both estrogen and testosterone.

The Other Female Hormones: Progesterone and Testosterone

Progesterone. Progesterone is also involved in mood and memory, though in a different role. As a counterbalance to estrogen, progesterone can offset some of the mood-modifying, memory-enhancing abilities of estrogen.[22] While I treat many of my female patients with various forms of progestins, primarily to prevent uterine cancer, I feel it is important to use the lowest dose possible that still protects the uterus. I also use natural micronized preparations, and have begun to use vaginal progesterone suppositories, to limit the potentially negative effects of progesterone on mood and cognitive function. ("Micronized" means that the particles are finely ground to improve absorption.) The vaginal method of delivering progesterone results in the hormone acting more on the uterus than in the brain. This allows progesterone to protect against cancer but results in fewer unpleasant side effects. In my view, these important modifications are going to become more popular as clinicians focus on the psychological effects of the hormones they give their patients. We must take care to respect and treat our patients' minds, just as we do their bones, breasts, uteri, and cardiovascular systems.

Testosterone. Testosterone is thought of as the male hormone, but as noted in the introduction, women produce testosterone, too, and it has an impact on their brains as well. (The opposite is also true; men synthesize estrogen and it plays a role in their brains.) Testosterone influences areas of the brain that regulate sexual responsiveness and cognitive functions, which include learning and memory. I first observed the cognitive effect of the male hormone when I treated a fourteen-year-old boy named Nick, who was born without testicles. The testes are the major source of testosterone production in men, and without them, Nick never experienced puberty. I put Nick on a regimen of testosterone injections every three weeks, which restored his normal pubescent development. With a gleam in his eye, he told me about his sudden capacity for erections. But something unexpected was also happening to Nick. One day, he asked me whether the testosterone injections might be helping him concentrate better in school. Nick noticed that his grades got better during the first two weeks of each treatment cycle, slumped during the third week, and then improved again after the next injection. When I shortened his injection cycle from three- to two-week intervals Nick was able to keep his schoolwork on an even keel.

Fifteen years later, I carried out an experiment in normal college-age males to explore whether there was a true hormonal basis for Nick's experience. I found that infusing testosterone into the bloodstream of test subjects, starting in the morning, improved the students' ability to complete a series of serial subtraction tasks in the afternoon.[23] This experiment and others proved that Nick had been right. The fourteen-year-old-boy had made a groundbreaking observation: Testosterone improves learning. Compared with our knowledge about estrogen, less is known about the effects of testosterone on neurotransmitters and their receptors, but there is no doubt that its multiple actions have a profound influence on sexuality and cognition.

Some of testosterone's impact on the brain may require that it be chemically converted to estrogen. In that case, the hormone acting on the brain is not testosterone at all but estrogen. In other cases, certain effects of testosterone appear to be related to specific actions of the hormone itself. Using a synthetic testosterone-like drug called mesterolone we were able to demonstrate an antidepressant effect in depressed men equal to that produced by a tricyclic antidepressant. Mesterolone showed an advantage over the antidepressant in that it was associated with fewer side effects such as a dry mouth or rapid pulse. None of the unpleasant sexual side effects seen with the antidepressants were observed with the hormone. Mesterolone cannot be converted by the body to estrogen, making its antidepressant effects independent of estrogen.[24]

Testosterone is a powerful treatment for sexual dysfunction in both men and women. In my clinical practice, I have found that low-dose testosterone is a key player in hormone modulation, not only to revitalize sexuality but also to stem the tide of depression and cognitive impairment. Higher doses of testosterone may also be extremely useful for men as they age, with positive results in terms of muscle strength, sexual vitality, sexual function, and mood.

Thyroid Hormone. The role of thyroid hormone in mental well-being has been vastly underrated. Put simply, if your thyroid is in trouble, all your endocrine glands are in trouble, since thyroid hormone regulates metabolism for the whole body. Women with hypothyroidism—an underactive thyroid with an inadequate output of thyroid hormone—typically have low estrogen levels. They also frequently suffer a complex of symptoms including depression, irritability, memory lapses, and cognitive fuzziness.[25] Whether low estrogen causes hypothyroidism or hypothyroidism causes low estrogen remains a chicken-or-the-egg question. But the two are linked, and there is a good chance that low thyroid hormone levels are both a cause and effect of low estrogen.[26] Either way, these women, most of whom have a form of autoimmune thyroiditis, often require treatment with both thyroxine and estrogen.

Why is thyroid hormone so important? In the brain, thyroid appears to augment neurotransmission. In my practice, the high incidence of low or borderline thyroid levels is such that I prescribe thyroxine to 10 to 15 percent of my patients. These are largely women who report mood or cognitive disorders that badly hamper their quality of life, and who experience major improvements with hormone modulation.

Thyroid hormones can also have pervasive effects on brain function. Consider, for example, the relationship between thyroid abnormalities and certain psychiatric disorders. A number of studies suggest that treatment with thyroid hormone benefits a certain percentage of nonresponders to antidepressants.[27] Patients with underactive thyroids frequently have a particularly virulent type of manic-depression known as rapid-cycling bipolar disorder.[28] Treatment with thyroid hormone results in remission of the rapid mood shifts characteristic of these patients.

Julie was a patient of mine who demonstrated the relationship between thyroid hormone and ovarian function, and the combined effect that estrogen and thyroid hormone can have on brain function. Her gynecologist told Julie, then thirty-two, that she was in early menopause, so he prescribed estrogen. The estrogen was helpful, but it would not be sufficient for Julie. I suspected a thyroid problem, and tests confirmed that she had an underactive thyroid—which no doctor had previously diagnosed. Her mood problems, lethargy, and sexual dysfunction resulted not only from premature menopause but also from autoimmune thyroiditis, which was causing her low output of thyroid hormone. Once her condition was treated with thyroid hormone, Julie's ovarian function returned to normal. I added a small daily dose of testosterone to restore her sex drive, and today she is mentally healthy and sexually vibrant. Julie's case exemplifies multileveled hormone modulation—the need to correct not one but several interrelated imbalances.

The Hormone Revolution: New Treatments for the Mind

Over the past decade, scores of studies have demonstrated the myriad ways that hormones regulate brain cell activity. The basic brain studies have been accompanied by clinical research demonstrating that estrogen, testosterone, progesterone, thyroid hormone, and other hormones help to regulate memory, concentration, learning, and behavior, not to mention sexual function and desire. We can no longer have any doubt that hormonal imbalances cause our patients to suffer a spectrum of cognitive, mood, and/or sexual problems.

Our brain cells are so exquisitely sensitive to the power of endocrine secretions that 100 picograms (a picogram is a trillionth of a gram) of estrogen in the blood is all these cells need to function at peak efficiency. As clinicians in the field of psychoneuroendocrinology—which links the mind, brain, and hormone systems—it is our job to be certain that our patients' brain cells are exposed to these picograms. If the brain is deprived of this hormonal immersion the results may include depression, moodiness, irritability, anxiety, anger, and fatigue. That is only the short list. The longer list includes decreased sex drive, inability to reach orgasm, cognitive confusion, and lapses in memory. We hear these complaints from our friends on a daily basis and are bombarded by television and newsprint with stories illustrating how common these ailments are, especially among women approaching or beyond menopause.

Given the research on hormones and the brain, it should no longer surprise us that declining estrogen and testosterone can have major emotional and cognitive consequences. Consider again that substantial decline in both of these hormones during the perimenopause, and the rather sudden 75 to 80 percent plunge in the months after the onset of menopause. Estrogen production falls even more dramatically and suddenly after childbirth. The massive hormonal decline in estrogen and progesterone levels within twenty-four hours after delivery is believed to be a major trigger for postpartum depression and psychosis in patients predisposed to these conditions. Levels of estrogen plummet from thousands of picograms during pregnancy, to 100 picograms or less postpartum. During the last half of the menstrual cycle, a five- to sixfold drop in estrogen levels occurs over two to three days. This decline, which is associated with the premenstrual rise and fall in progesterone, is sufficient to spur anxiety, mood swings, and other psychological symptoms of premenstrual syndrome (PMS) in susceptible women.

Is it a coincidence that women experience mood disorders far more frequently during the premenstrual phase of their cycle, after childbirth, and during and after the menopausal transition? Is it a meaningless statistic that twice as many women as men are diagnosed with clinical depression? We now have a wealth of evidence suggesting that these are no anomalies. The fall in estrogen and testosterone levels during these periods is a pivotal factor. (Certainly, depression is a complex condition—genes, upbringing, stress, and life events are usually involved. But given what we now know, ignoring the hormone connection is scientifically unjustified.) Indeed, the clinical successes of hormone modulation in the treatment of mood disorders are further proof that hormones are vital contributors to psychological health.

Beyond depression, the memory lapses and cognitive impairments common among peri- and postmenopausal women are also highly responsive to estrogen-based treatments.

The mental health indications for hormone modulation don't stop with mood and cognitive impairments. Estrogen can alleviate symptoms of manic-depressive and schizoaffective disorders, and other types of psychosis that can occur during the first week after giving birth. Estrogen depletion has also been implicated in schizophrenia, the severe psychiatric disorder characterized by hallucinations and detachment from reality. While the majority of schizophrenic patients first become sick in early adulthood, there is a second peak in schizophrenia onset in women—but not men—after age forty. Studies suggest that estrogen loss may be implicated, at least in part. Severe premenstrual mood swings, the scourge of millions of women, often respond dramatically to estrogen therapy.

The recent brain studies and clinical trials place us on the cusp of a revolution in the treatment of disorders of the mind. It's a hormone revolution, and while more research must be conducted on proper indications and risk management, the medical world is just beginning to consider the vast potential of hormones for treating the mind. We can only hope that the transformation of lab findings to clinical treatments will be both swift and sure, and that this book will help both patients and their doctors to advance the cause of safe, effective hormone modulation for the most difficult-to-treat maladies of mood, emotion, and sexuality.

2

Reclaiming Emotional Balance

DEPRESSION AND MOOD

PAULA'S STORY

Paula, a hardworking office manager, came to me in a state of severe emotional and physical pain. Previously, she'd experienced only one brief episode of depression. Five years earlier, her teenage daughter, Ann, was acting out, as some teenage daughters do. But Ann's behavior was so out-of-control that it was causing strains in the family. She and her husband Bob were constantly on edge, and finally, Paula saw a doctor who prescribed Prozac. The Prozac was effective, and in time, her daughter began seeing a counselor who helped her to curb her acting-out. As Ann got better, Paula became less stressed out, and she was able to wean off Prozac. But now, Paula, at fifty-three years old, was in the throes of a serious depression. She was constantly on the verge of tears, and her sex drive was nonexistent. There was no obvious cause in her daily life, no major disruption in her work or family life.

Paula was not only depressed; she was experiencing pain in her joints and muscles that interfered with her work. Her doctor thought she had fibromyalgia, a condition in which persistent aching pains affect the neck, back, hips, elbows, and knees. The causes are not fully understood, but many physicians—myself included—have noted that it frequently occurs around menopause. Paula's doctor explained that antidepressants had shown some benefit for fibromyalgia, so he prescribed one called

Effexor. He explained that the drug would take at least one month to start working.

But two months later, Paula was no less depressed, and her aches and pains had not improved, either. Moreover, she was having even more trouble with her sex life. (This could have been partly a side effect of the drug.) Vaginal dryness was making sex with her husband painful and this difficulty prompted Paula to wonder whether menopause was behind her complex of symptoms. She hadn't had a period in over a year. Until then, she doubted whether menopause was a factor, since she hadn't experienced the usual hot flashes and night sweats that her menopausal friends described so vividly. (She couldn't have known that menopause can trigger symptoms such as depression and fibromyalgia, and that these problems can occur unaccompanied by the more typical hot flashes.) Paula remained on her Effexor for another month, but when she still wasn't getting better, she came to me for a consultation.

I listened carefully as Paula told me her story, and I tested her blood for hormone levels. As I suspected, her estrogen levels were extremely low, and her follicle stimulating hormone (FSH) was highly elevated—both signs of menopause. Her monoamine oxidase (MAO) level, measured in blood platelets, was also very high, which can be linked both to menopause and depression. Finally, her testosterone blood levels were found to be dramatically low. I was convinced that these low hormone levels, common in women at or near menopause, were behind Paula's lack of response to antidepressants.

I recommended that she remain on Effexor, but I added an oral estrogen to her regimen. After one month of treatment, Paula's depression was markedly better, and the debilitating pains of fibromyalgia had largely disappeared. She'd been on Effexor for three months without improvement; once she added estrogen, much of her suffering was finally relieved. Vaginal dryness was no longer a problem, though she still did not have her normal sex drive. Given her low testosterone levels, I felt that the addition of a low dose of an oral testosterone, methyl testosterone, was justified. After a short while on the testosterone/estrogen combination, her depression had totally lifted, and her fibromyalgia was a thing of the past, and her sex drive improved.

Paula's story is common, not only in my practice but among women undergoing hormonal flux, whether during the premenstrual phase of their cycle, after giving birth, or during the perimenopausal years. During these times, a surprising number of women are not only beset by mood disorders, they are *not* made well by the medications prescribed by their doctors and psychiatrists. Reports of the incidence of treatment-resistant depression vary. According to leading psychiatrists at Yale University, as many as 40 to 50 percent of depressed patients "do not experience a timely remission" with antidepressant treatment. More conservative estimates suggest that 15 to 30 percent of patients don't get well on medication. Many of these patients go through several trials of different drugs. Some get better but a sizable group does not. Others feel better for months until the effect wears off, while still others cannot tolerate the side effects of the medication.

Based on my clinical practice, I believe that many depressed patients who don't respond to standard treatments are women with hormonal imbalances. Unfortunately, as leaders in psychiatry have acknowledged, we have little data on people with treatment-resistant depression, but we know it's a major problem. I am convinced, not only from my own practice but also from the hundreds of studies on hormones and the brain, that hormone modulation is the missing piece for many suffering patients—the one that finally allows them to benefit from antidepressants. Estrogen and testosterone were the missing pieces for Paula, who finally got well after months of psychological and physical distress.

When psychiatrists and other doctors are presented with patients like Paula, they usually prescribe an antidepressant or tranquilizer. When the symptoms of perimenopause are primarily emotional and not physical, the possibility of hormonal intervention is typically overlooked. More broadly, the medical profession has yet to fully recognize the dimensions of hormone-associated mood disorders.

In this chapter, I will explain the role of hormones in mood disorders. I'll return again to the story behind the story, those molecular locks—"receptors"—that receive the neurotransmitter "keys." Antidepressants boost the number and activity of those "keys," but in treatment-resistant patients, there can be a serious problem with the "locks." They lack a sufficient number or quality of receptors in the brain, which means that all the neurotransmitters in the world will not help. I tried to explain this in lay language to one patient, who had failed to respond to many different antidepressants. She summed it up better than I could: "You mean, you can have a hundred different keys but if you don't have locks that fit right, you'll never get well."

Hormones—estrogen increases the numbers and sensitivity of those molecular locks, allowing for the key-in-lock connections that help people get well. It may not be the problem in *all* cases of hard-to-treat mood disorder, but it *is* the problem for many women, and some men. In my view, every person with persistent depression should have a hormonal profile to determine whether hormone imbalances are causing or contributing to their condition. Some doctors may not yet believe this is justified, but read ahead and decide for yourself whether you believe that hormones are a factor in your suffering. As doctors and patients begin to take this evidence seriously, I believe that hormone modulation will become a standard therapy for vast numbers of men and women in the coming era of medicine. Because women are more prone to depression, they may especially benefit from this therapy.

WOMEN, DEPRESSION, AND HORMONES

We know from reams of research that women experience twice as many depressive episodes as men in their lifetimes, and are two times more likely to be hospitalized for depression.[1] This striking gender difference has been observed in countries worldwide, though the reason has not been fully established. We know that it's not because women seek professional help more frequently than men. It isn't related to any bias on the part of physicians to more readily diagnose depression in women. While social, economic, cultural, and genetic factors may be involved, none of these influences adequately explain why women suffer so much more depression. Nor does research prove that women are more frequently depressed than men simply because they are more stressed out.

Can this all be explained by hormones? We can't yet say, but many facts and observations lead me to believe that the disproportionate incidence of depression in women is partly, if not largely, due to hormonal factors. The events in women's lives most commonly associated with depressive episodes are hormonal. Specifically, precipitous drops in estrogen levels at various phases of the life cycle can have profound effects on the mind:

- *Premenstrual:* Just prior to menstrual flow, estrogen levels decline rapidly. Thirty-five percent of women have moderate premenstrual physical and depressive symptoms, and 3 percent suffer severe, incapacitating symptoms. Prozac and other antidepressants are being prescribed with increasing frequency for the mood disorders associated with PMS.[2]
- *Postpartum:* Estrogen levels plummet dramatically in the "postpartum" period following pregnancy, when there can be a rapid decline

in estrogen to levels that are 100 times less than during pregnancy. Three to ten days following delivery, 50 to 70 percent of women experience a mild, transient mood disorder. Within the first three months, 10 to 15 percent suffer a major depressive disorder, and 0.1 to 0.2 percent of women experience a complete break with reality—postpartum psychosis. The horror of psychosis can have a violent face and some psychotic postpartum mothers have killed their babies.[3]

A study at the Helsinki City Hospital in Finland found women with a postpartum psychosis responded positively to estrogen treatment.[4] Women with low blood estrogen levels had reversal of their psychosis rapidly on treatment, and psychotic symptoms diminished significantly during the first week of estrogen treatment. By the end of the second week all patients became symptom free. One patient who discontinued hormonal therapy suffered a recurrence of her psychosis.

Another study at the NIH determined that women with a history of postpartum depression are uniquely sensitive to the mood destabilizing effects of estrogen and progesterone.[5] Eight women with and eight women without a history of postpartum depression were administered Lupron, which suppresses estrogen and progesterone secretion by the ovary. Each group was then placed on high doses of estrogen and progesterone to simulate pregnancy levels. Under double-blind conditions after eight weeks both estrogen and progesterone were withdrawn, imitating the hormonal conditions found postpartum. More than 60 percent of the women with a history of postpartum depression developed mood symptoms, whereas none of the women without such a history became depressed.

- *Menopause and Perimenopause:* The most widely recognized decline in estrogen is at the time of menopause, when menstruation ceases and estrogen levels fall 75 to 80 percent. This drop may cause psychological symptoms in women whose genes or previous depressive episodes make them susceptible. But prior to this marked decline, a 40 to 50 percent decline in estrogen occurs between the ages of twenty and fifty.[6] By the time a woman is into her forties, estrogen levels may be low enough to affect mood, memory, and sexuality. This period is commonly known as the perimenopause. Periods become irregular and moods swings can be extreme. Studies suggest an increased incidence of mood disorder during the perimenopausal years.[7]

As I've emphasized, women's estrogen levels can start to decline when women are in their twenties, though the time frame known as perimenopause usually begins sometime in their forties. The initial slow decline in ovarian function and then the precipitous drop at menopause puts women at greater risk for depression. Indeed, the National Institute of Mental Health (NIMH) suggests that perimenopause is one of the times when women are most vulnerable to mood disorders. This was confirmed by a recent study of 477 women, in which investigators found a high prevalence of mood disorders *during the three years surrounding menopause.* During this time, 30 percent of the sample scored high on a test for major depressive or primary anxiety disorder.[8]

Studies carried out in menopausal clinics and gynecologists' offices suggest a link between declining estrogen and mood disorders. Many of these studies support the view that women in perimenopause are more emotionally labile—susceptible to mood swings and disorders—than women well past menopause. In one such clinic, at the Royal Edinburgh Hospital in Britain, 45 percent of the patients were categorized as clinically depressed. Eighty-three percent of the depressed patients had experienced a previous depressive episode, compared to only 33 percent of the women who were not currently depressed. A clear peak of depressive illness was seen in the four years on either side of the last menstrual period. Thirty-five percent of all patients with past or current depressive illness experienced their first depressive episode in this perimenopausal period.

The clear implication from the British study is that estrogen loss alone does not explain depressive episodes occurring at the perimenopause and menopause. A prior episode of depression is a major factor predisposing to the development of a depression as estrogen levels drop. Put differently, women already susceptible to depression for various other reasons—i.e., genes, life events, and stress—are more vulnerable to depressive episodes when their estrogen levels fluctuate and decline.

To those who feel that discussing hormones in the context of mental health is somehow inappropriate—whether on political or biological grounds—I can only say that neglecting the role of hormones would be a willful evasion of scientific truth. Even worse, it would prevent us from developing and refining treatments that can help women who are suffering with symptoms that only serve to degrade their quality of life.

The truth about mood disorders during the menopausal transition should not be used to advance shopworn ideas about women becoming unreliable or incapacitated as they age. In her book, *The Pause,* psychologist Lonnie Barbach, Ph.D., who's on the clinical faculty at the University of California Medical School in San Francisco, addressed this issue:

> There are those who, in an attempt to emphasize the fundamental equality of the sexes, would minimize the effect of hormones upon our emotional equilibrium, and at times the feminist in me is tempted to say that in this regard we are no different from men. But in truth we are. And while many women remain emotionally unaffected by the change in their hormonal balance, others are affected to varying degrees. Certainly this does not mean that we are rendered incompetent at this time in our lives, and denying us opportunities because of potential vulnerability to hormone fluctuations is unwise, unfair, and unjust.[9]

Both women *and* men experience hormonal fluctuations that can affect their thinking, behavior, moods, and sexuality. While the genders differ in terms of distinct hormonal factors that cause distinct symptoms, there is no medical reason or rational justification for sexist attitudes. Testosterone imbalances may contribute to excessive male aggressiveness, but no one uses this fact to advance antimale attitudes. We should be able to explore how hormones affect men and women differently without falling back on long-discarded biases.

IS ESTROGEN AN ANTIDEPRESSANT? THE WORCESTER STATE HOSPITAL EXPERIENCE

My first indication that hormones might be related to depression grew out of my research with my collaborators, Dr. William Vogel and Dr. Donald Broverman. As mentioned in chapter 1, we focused on an enzyme active in the brain, monoamine oxidase (MAO) and its relationship to both estrogen and depression. We knew that MAO enzymes acted like Pac-men chewing up neurotransmitters, including norepinephrine and serotonin. Given the importance of these neurotransmitters in maintaining balanced mood, too much MAO in the brain can result in mood disorders, most notably depression. In one of our earliest studies, we measured MAO activity in the blood of clinically depressed women, and found—as we had hypothesized—that their levels were high. At the same time, we discovered that we could lower MAO activity in these women by administering estrogen. Not only did their MAO drop, but their mood states improved markedly.

With these findings in hand, we successfully applied to the NIH for support of a clinical trial of estrogen as an antidepressant. We chose the sickest patients on the wards of the Worcester State Hospital, severely depressed women, both pre- and postmenopausal, who had not responded to any earlier psychopharmacological treatment. Sixty-three percent had made serious attempts at suicide and all had a long history—at least two years—of having tried antidepressant and/or tranquilizing medications. Almost half had received electroshock therapy. Six women in the estrogen-treatment group and four in the control group had psychotic depressions. In the language of psychiatry, these patients had "primary, recurrent, unipolar, incapacitating, major depressive disorders."[10]

Twenty-three of these women were randomly assigned to our estrogen-treatment group, and seventeen to a placebo control group. (They would receive sugar pills instead of estrogen.) Given the severity and longevity of their illnesses, there was little or no expectation that these forty women would spontaneously get well. As my colleagues and I talked with these patients, we were often struck by their desperate desire for relief from suffering. Many of them viewed entrance into this experimental research project as a last chance to get well after all else had failed.

This was a placebo-controlled, double-blind study, in which neither the patient nor the prescribing doctors knew whether the patient was receiving the drug or the sugar pill. Our previous research made it clear that these patients would likely require large doses of estrogen to have sufficient impact on the brain. So we gave each estrogen-treated patient 5 mg daily, increasing the dosage in 5 mg steps each week that she failed to show improvement. The maximum daily dosage was set at 25 mg, which only half the patients needed. Premenopausal women also received progestins for five days each month in a cyclic fashion.

Patients' mood states were assessed regularly throughout treatment using the standard Hamilton Scale of Depression. After four months, we discovered that the women receiving high-dose estrogen, as a group, experienced significant reductions in the symptoms of depression, while placebo-treated patients showed no change or became worse. Specifically, for about 25 percent of the estrogen-treated women, the depression scores *returned to normal.* For 60 percent, the depression scores were significantly reduced. In short, some patients improved remarkably, while most had modest but statistically meaningful improvements in their conditions. Given how severely ill these patients were at the outset, our findings could only be considered striking. The study, published in 1979 in the *Archives of General Psychiatry,* is frequently cited as one of the earliest pieces of convincing evidence that estrogen has antidepressant properties.

Over the four months 47 percent of the estrogen-treated patients experienced a positive change. Only two of the twenty-three treated patients became more depressed (8.7 percent), while eight of the seventeen placebo patients—a sizable 47 percent—experienced a worsening of their condition.

We also found, clinically, that many of the estrogen-treated patients in the study not only got better, they were able to consolidate and extend their improvements well beyond the four months.

As a scientist, I was gratified by the changes in the Hamilton ratings, since this enabled us to scientifically establish the antidepressant properties of estrogen. But the real reward came as I watched some of the women leave the hospital transformed. Families that had been shattered were restored. Mothers were able to resume care of their children and wives began to reestablish loving relationships with their husbands.

Helene, who'd been hospitalized for fifteen years, was one of the patients who'd suffered from psychotic depression. Her husband would frequently visit her in the hospital prior to her estrogen therapy. Helene was immobilized, and subject to occasional hallucinations; she simply could not function. She had been on antidepressants, including MAO inhibitors, and received electroshock treatments. Nothing worked. The couple had two young children, and I recall the husband's sense of sorrow when he would leave, unable to take Helene home. On the few occasions when Helene was cleared to go home, she'd have to return within a few days.

Estrogen therapy triggered a stunning turnaround in Helene's condition. Her despair lifted, her hallucinations ceased, and her energy improved. She was able to return home to her husband and children. Helene and her husband visit me regularly, enabling me to keep tabs on her condition and make certain that she does not suffer a relapse. I can report that Helene's spirits remain good, and that she's grateful for the twenty good years she's had with her family.

The very high doses of estrogen used in the Worcester State Hospital study would not be used today. But we established the principle that estrogen may be a missing link in the treatment of severe, treatment-resistant depression. We also demonstrated higher-than-usual doses of the hormone may be required. I should add that we have many new and more effective antidepressants since our study was carried out, yet many women still fail to respond or become resistant to treatment.

HORMONES FOR DEPRESSION: WHAT WE KNOW

The publication of our findings sparked a great deal of excitement about estrogen treatment for severe depression, but it didn't last long. Over the next few years, concerns that estrogen caused uterine cancer put a definite damper on efforts to study estrogen for depression. To compress a lengthy and complicated history, during the early 1980s it was found that adding progesterone to estrogen-replacement regimens eliminated any increased risk of uterine cancer in menopausal women on estrogen. This reopened the door for research on hormones for depression, and a considerable body of evidence has amassed.

I'll summarize this evidence in concise form. In 1997, Drs. Julianne Zweifel and William O'Brien of Bowling Green State University published the findings from their overview of twenty-six clinical studies of hormone treatments for depression.[11] When the studies, which involved 1,226 patients, were pulled together and analyzed, Zweifel and O'Brien concluded that hormone therapy was "effective in reducing depressed mood among menopausal women." The overall "effect size" was "moderate to large"—a statistic indicating that the average patient receiving hormones had lower levels of depression than 76 percent of the patients in control groups.

Zweifel's and O'Brien's study provides fresh and convincing evidence that hormone modulation with estrogen, or estrogen plus testosterone, is a viable, effective treatment for depression among peri- or postmenopausal women. Because it combines twenty-six well-designed, peer-reviewed studies into one pooled result, their findings must be taken seriously.

Pioneering researcher Barbara Sherwin, Ph.D., of McGill University in Montreal, conducted several of the most important studies included in the Bowling Green analysis. In two experiments, Dr. Sherwin investigated hormone treatments for women in surgical menopause who had undergone a total hysterectomy with removal of their ovaries. In both studies, women receiving hormones—no matter what regimen—had lower depression scores than those who got placebo injections. Moreover, during the month when all the women were put on placebos, their depression levels shot up. Sherwin took this a step farther. She drew blood samples from the women, and during the placebo phase their blood estrogen and testosterone levels took a dive—along with their moods.

The results of her study clearly indicate a direct relationship between more positive moods and higher blood levels of estrogen and testosterone.

Overall, the evidence indicates that estrogen-based hormone treatments prevent or reduce depression in peri- and postmenopausal women, whether their loss of ovarian function is natural or surgical. Now let's explore why and how these hormone therapies work, how they can best be applied, and what it all means for the treatment of mood disorders.

HORMONES AND THE BIOCHEMICAL BASIS OF DEPRESSION

To understand the remarkable power of hormones to influence emotional well-being, it will help to briefly review the biochemical basis of depression. What we discovered is that estrogen acts as a positive facilitator at many steps in the normal brain processes that regulate mood. Progesterone and testosterone play important roles, too. Neurotransmission is the pivotal process which when it functions properly creates harmony between all sections of the brain. When this happens, we avoid the disharmony that leads to depression and other mood disorders.

Over the past two decades, neuroscientists and psychiatrists have made monumental advances in their knowledge of the chemical basis of depression. We now believe that three neurotransmitters—norepinephrine, serotonin, and dopamine—are primary regulators of mood in the brain. Mood disorders, including depression and manic depression, result from an imbalance or deficiency of these neurotransmitters; a disorder in their metabolism; or a failure of these molecules to properly make their connections with receptors on neighboring neurons. Much of our understanding has grown from research investigating the actions of antidepressant medications.

The Role of Estrogen

Over the past decade, evidence has built that estrogen and other sex hormones are directly involved in processes of neurotransmission involving serotonin, norepinephrine, and dopamine. I'll concentrate here on the estrogen-serotonin connection, the most interesting and arguably most important relationship between a hormone and a class of brain chemicals that determine our moods. Remember, for balanced moods we need enough serotonin. We also need serotonin receptors—the keyholes—that are sensitive, that bind properly to serotonin, and that are sufficient in number—neither too few nor way too many. Consider the following effects of estrogen on serotonin neurotransmission, which I discussed in chapter 1:

- *Estrogen inhibits the breakdown of serotonin by the MAO enzyme, thereby increasing the amount of brain serotonin*
- *Estrogen boosts brain serotonin synthesis*
- *Estrogen increases the number and sensitivity of serotonin receptors*

Dr. George Fink, a world-renowned neuroscientist and expert in mood disorders such as depression and schizophrenia, of the Brain Metabolism Unit at Edinburgh University in Scotland, gave estrogen to estrogen deficient female rats and found a pronounced increase in the density of serotonin receptors in the certain parts of the brain.[12] This receptor effect may suggest that sex differences in the incidence of depression and the antidepressant effects of estrogen may be due in part to the action of estrogen on serotonin receptors. Dr. Fink calls estrogen "nature's psychoprotectant."[13]

Estrogen and Treatment-Resistant Depression

We are learning, to borrow the headline of a recent *New York Times* article, that "some still despair in Prozac Nation." Treatment-resistant depression is among the most difficult problems faced by psychiatrists and other physicians. As I mentioned, about 15 to 30 percent of patients don't respond to their antidepressants; one estimate puts this figure closer to 40 to 50 percent. Some patients finally respond after many trials with different drugs, but others—no one knows exactly what percentage—are not helped. Still others get better for a time until the medication stops working.

The *New York Times* article, which ran on the front page of the science section, shed light on treatment-resistant depression. Experts quoted in the article admitted that there has been little systematic research in this area, though they acknowledged the seriousness of the problem—and the dearth of solutions. David Kupfer, chairman of psychiatry at the University of Pittsburgh School of Medicine, admitted that many clinicians, faced with patients who do not improve, "tend to throw up their hands."

As I've argued, estrogen and other hormones are often the missing pieces for people with treatment-resistant depression. The reason is that antidepressants boost the neurotransmitter "keys," but unless the person has the proper amount and quality of receptor "locks," the treatment simply won't work. Estrogen has a powerful effect on receptors for serotonin and norepinephrine, and so do testosterone and thyroid hormone. Women with plummeting levels of estrogen whether associated with the premenstrual phase of the menstrual cycle (usually seven to ten days before the flow begins), the postpartum period, perimenopause, or menopause, may be vulnerable to treatment-resistant depression. (Other factors—mainly genes and stress—make them susceptible.) That's why I run a complete battery of hormone tests as soon as a patient comes to me and says, "I'm depressed and no treatment is working."

What do I find on these tests? With alarming frequency, women with treatment-resistant depression have very low estrogen and testosterone levels. Of course, among postmenopausal women this is no surprise, but I often find relatively low estrogen and testosterone levels among women in their twenties, thirties, and forties. Many are perimenopausal and don't know it, because they haven't had any hot flashes. Other women have subclinical or previously undetected hypothyroidism. The bottom line is that women with refractory depression often have an underlying hormonal imbalance.

Research is now being conducted to determine whether hormone treatment is one answer to treatment-resistant depression. Already, there is strong preliminary evidence that estrogen can help those who have not responded to antidepressants. A recent report of a multiple medical center study describes the results of research underway on 658 women over age sixty with major depression.[14] The patients were all taking Prozac, and a mere 32 percent had a positive response, as compared to 18 percent on a placebo. This low positive response to Prozac is much less than the usual 70 to 80 percent. But when the researchers identified women who'd been receiving estrogen replacement, they had a dramatically better response to Prozac than the women not receiving estrogen.

In another report on women over age sixty, 72 patients received ERT and 286 did not. In these two groups of women an antidepressant response to treatment with Prozac (fluoxetine) or a placebo was significantly different. Women not on estrogen receiving Prozac did not show any benefit significantly greater than the placebo-treated women not on

ERT. Estrogen use appeared to augment Prozac response lowering the Hamilton Depression Scale score by 40 percent versus 17 percent for patients who received the placebo.[15]

Treatment-resistant depression is more common among the elderly but older depressed women are not the only ones who may be helped by hormones. A study at Columbia University in New York City examined a similar effect of testosterone replacement on depressed middle-aged men with low testosterone levels. These men had failed to respond to an adequate trial with antidepressants similar to Prozac. When testosterone was added to the antidepressant regimen, the researchers documented a dramatic rate of recovery from major depression. When testosterone was withdrawn and a placebo was substituted, 75 percent of the subjects had recurrence of depressive symptoms.[16]

Obviously, more research is needed to confirm these preliminary findings, but the results to date are encouraging. They bolster my strong clinical impression that both estrogen and testosterone can be used successfully to augment antidepressant therapy. I have had scores of patients, struggling for years with refractory depression, finally respond to their medications when hormones are thoughtfully and judiciously added to their treatment regimens. One such patient was Sharon.

SHARON'S STORY: OVERCOMING TREATMENT-RESISTANT DEPRESSION

Even as a young child, Sharon had been beset by anxiety and depression. Her condition was so bad that she began seeing a psychotherapist at age eight. She had a family history of alcoholism, anxiety, and agoraphobia. In her teens, she was given oral contraceptives—a hormonal combination that can destabilize moods—and she became severely anxious and depressed. "I've always had a problem with major depression," she said. "But the pill multiplied everything by a thousand." Over the next ten years, her condition, which was so severe it involved suicidal thoughts, was treated with a wide variety of antidepressants and mood stabilizers, with only temporary benefits.

After Sharon got married, she had two children, a daughter and a son. During both pregnancies, her mood was so much better that she was able to get off antidepressants. (Estrogen levels are sky high during pregnancy.) Giving birth, however, was not to be followed by joy. After both pregnancies, she suffered severe postpartum depression. (Maternal estrogen levels *plummet* after birth.) The first time around, Sharon responded well to treatment with Zoloft. But after the birth of her second child, her postpartum depression did not respond to Zoloft. Sharon also had a history of migraine headaches, which worsened after her first delivery. When she developed postpartum depression after the second pregnancy, the migraine headaches became even more frequent and debilitating.

Sharon came to see me several years after her second bout of postpartum depression, because she was only getting worse. Then thirty-five years old, she complained of constant agitation, depression, and deadening fatigue. "I can hardly move," she said. "The house is a mess, and everything overwhelms me. All I can do is retreat to the bedroom." Sharon had even cut back on her work as an interior designer. She was then taking two antidepressants, with no benefit. Her physician had treated her aggressively, trying many medications and raising the dosages. "I felt like nothing was getting through," says Sharon in retrospect. "I thought, how can it be that nothing is touching this?"

I told Sharon what I suspected: that hormone imbalances were the crux of her problem. She experienced my remarks as validating—even liberating. In her mind and heart, she had believed this all along. Her hormone test results confirmed both our suspicions: She had relatively low levels of estrogen and markedly elevated platelet MAO levels. I thought that low estrogen was the probable cause of her poor response to antidepressants. To be effective, Zoloft requires functioning serotonin receptors, and in light of her estrogen deficiency, I doubted that Sharon's receptors were functioning properly. Early in her treatment, I ordered a brain Single Photon Emission Computerized Tomography (SPECT) scan, which showed a slightly but noticeably decreased blood flow to several areas of Sharon's brain. This is a frequent finding in depressed individuals, and it helped confirm my hunch that her brain was not getting the hormonal stimulation it needed for adequate blood flow.

I started Sharon on oral estrogen while she remained on the same antidepressants. One month later, she was still somewhat depressed, but not quite as despairing, and she was functioning better. By the end of the second month, the change had solidified: Her depression was lifting and her migraine headaches were far less frequent. Blood tests confirmed that her improvements were accompanied by an increase in blood estrogen levels.

Sharon also had low blood testosterone levels. In the hope of consolidating her gains, I prescribed testosterone. Though the initial dosage was low, Sharon became somewhat agitated on testosterone and I promptly reduced the amount. After about four months of treatment, I met again with Sharon. Now, without doubt, she was fully responding to her antidepressants. "I feel better than I have in years," she said. "I know there's some fine-tuning to do, but I'm finally branching out, living my life again." She began taking on new work assignments, and started playing the guitar and singing. We continue to meet, and we've tweaked her regimen a few times—changing antidepressants—but she remains free of her mood disorder.

One story such as Sharon's does not prove that hormones can cure treatment-resistant depression. But I have scores of similar cases in my clinical files, and the burgeoning research in psychoneuroendocrinology supports my view that patients like Sharon truly benefit from hormone modulation. While we still need those large-scale clinical trials, I believe that judiciously applied hormone treatments are justified in cases such as hers.

ESTROGEN AND ANXIETY

While depression is the most common mood-related complaint in peri- and postmenopausal women, it is frequently accompanied by anxiety. In fact, when I first see many female patients in their forties or fifties, the problem they see me about is anxiety. Their typical symptoms include a dry mouth, excess sweating, waking in the night short of breath with a pounding heart, nausea, or diarrhea. Estrogen treatment can often relieve these symptoms of anxiety as effectively as it can alleviate depression.

NORMA'S STORY

Norma was a perfect example of how estrogen can relieve anxiety. Two years before I first saw her, she was nearing menopause and having some minimal symptoms of hot flashes and night sweats. Her gynecologist started her on an oral contraceptive containing both estrogen and a progestin. After a year on this medication, Norma began to notice that she would wake up in the morning feeling anxious and slightly irritable. Occasionally, severe dizziness would sweep over her while she was walking. She began to have anxiety attacks that verged on panic attacks. The final straw that prompted Norma to seek additional treatment occurred at an afternoon luncheon. "The whole room started spinning," said Norma. "I felt like I was going to faint in the middle of this luncheon full of people. They wanted to call an ambulance."

Her physicians did not have answers. "They shrugged it off as perimenopausal symptoms." While there was some truth to this comment, the "shrugging off" was the problem. The oral contraceptive was causing more trouble than good, and what Norma needed was appropriate hormone balancing. When I told Norma that I thought her problem might be hormonal, her response was, "But I'm already receiving hormones. How could that be my problem?"

Norma's anxiety had two causes. First, the oral contraceptive she was taking every day combined estrogen and progesterone. This is not a happy combination for an anxious, depressed, menopausal woman. While the estrogen may calm her anxiety and elevate her mood, the progestins in this preparation can have the precise opposite effects. In Norma's case, the progestins were winning out. The second problem was her menopausal decline of estrogen. Given Norma's condition, the oral contraceptive was not an appropriate way to balance her hormonal profile. When I discontinued this pill and treated her with estrogen alone, her symptoms subsided but did not totally disappear. I suggested that she see a psychiatrist, who started her on a very low dose of Prozac. Within weeks, her anxiety had completely vanished.

"I feel great," she says today, two years later. "Once in a while I have a little dizziness, but I know it's the progesterone suppository and that it won't last long. Otherwise, nothing stops me."

Norma was one of many women who become either depressed or anxious when they are exposed to progestins on a daily basis. This can be an extremely serious problem since menopausal women on estrogen need to take a progestin to protect their endometrium (uterine lining) against cancer. In such women, exposure to even low doses of progestin can be devastating. I decided to treat Norma with natural micronized progesterone that is delivered by vaginal suppository, and taken for ten days each month. (More on this shortly.) On this regimen, Norma had few of the untoward psychological side effects of synthetic oral progestins.

The Role of Progesterone

Norma's story demonstrates a point that physicians and gynecologists have commonly observed—that women taking a combination of estrogen and progestin, either in a birth control pill or in hormone replacement, may have problems with mood. (This observation is often made when women are taking progesterone only part of the month, and it's obvious that their mood worsens during this time frame.) They may experience moodiness, weepiness, irritability, confusion, memory problems, and a host of other psychological symptoms. Many patients who consult with me are suffering from the negative effects of progesterone, and as a clinician I have devoted myself to finding healthy solutions to their problems.

How can a "natural" hormone cause such trouble? One reason that women on hormone replacement must take progesterone is that it offsets the stimulating effect of estrogen in the uterus. Unopposed estrogen (without progesterone) uterine stimulation can lead to cancerous changes. Progesterone reduces the increased uterine cancer risk associated with estrogen. In other words, progesterone can put the brakes on estrogen in certain parts of the body. It plays a similar role in the brain, and in people with well-balanced moods, progesterone does its job correctly. But too little estrogen and/or too much progesterone are a prescription for mood disorder. When a woman's estrogen levels are dropping, she doesn't need extra progesterone to put the brakes on estrogen in the brain.

The actions of progesterone that counter estrogen's brain stimulating activity are:

- Progesterone can *inhibit* estrogen receptors diminishing the effect of estrogen.
- Progesterone appears to have a dampening effect on brain receptors for serotonin and norepinephrine.
- Progesterone can also increase brain MAO activity, which hastens the breakdown of norepinephrine and serotonin.
- Progesterone can dampen women's sex drive and function competing with testosterone, a hormone that is essential to healthy sexuality in both women and men.
- Progesterone stimulates brain receptors for the amino acid gamma-aminobutyric acid (GABA), which is responsible for reducing anxiety. This effect can also induce sleepiness.

How are Progestins Administered?

Individualized hormone modulation for mood disorder requires careful attention to the forms and doses of progesterone, which often must be given along with estrogen. Current HRT typically employs either a continuous or cyclic form of synthetic progesterone (progestin) administration. In the cyclic regimen, estrogen is taken daily and progestins are added for ten to fourteen days each month. In the continuous regimen, a combination of estrogen and progestins is taken daily.

Two forms of continuous estrogen/progestin treatment are in wide usage today. One is the oral contraceptive, and the other is a form of noncontraceptive estrogen and progestin called Prempro, commonly used to treat menopausal symptoms. Many women find these continuous progestin approaches to be very convenient. The daily administration of progestin suppresses the growth of the uterine lining (endometrium), frequently to the point that women no longer are bothered by menstrual periods. Problems arise when some women, sensitive to continuous progestin, notice negative changes in their mood—usually depression or anxiety. I've had numerous menopausal women who complain of massive mood shifts and deepening depression while taking continuous combined estrogen/progestin. They frequently were referred by physicians to a psychiatrist for treatment with antidepressants. In these cases, antidepressants are not effective. Not until the combined estrogen/progestin regimen was discontinued did they begin to respond to their antidepressants.

SALLY'S STORY

Care must always be taken with the progesterone component of hormone modulation. Consider the case of Sally, whom I first saw when she was fifty years old. Her menstrual periods had become irregular, and she was waking up in the middle of the night, unable to sleep. Finally, after hours of tossing and turning, she would fall asleep, which only made it harder to get up in the morning. "I wake up more exhausted than when I went to bed," she said. "I'm so tired I can't think. There is no way I can function like this. My energy is nonexistent and my head aches all the time."

In the midst of divorce proceedings, Sally was crying frequently and had been depressed for the previous six months. Four months before I saw her, she had an episode of anxiety, and her doctor had given her Valium. "I feel like I'm hanging by my fingernails," she told me.

Her story exemplifies the mood disorders and related physical symptoms that can have a devastating effect on the lives of peri- and postmenopausal women. Sally's blood testing revealed that she had a low blood estrogen level and a high follicle stimulating hormone (FSH) level, compatible with the onset of menopause. I started Sally on Premarin, the most frequently prescribed form of estrogen. A month later, at her next visit, she was far less depressed and anxious, while her insomnia had lessened and her energy was vastly improved.

But in addition to estrogen, I put Sally on natural micronized progesterone during the first ten days of each month, in order to protect her endometrium—the lining of the uterus—from being overstimulated by the estrogen. Treatment with a progesterone-like medication is necessary in order to prevent the development of uterine cancer in women taking estrogen. With this treatment, women on estrogen are no more prone to uterine cancer than women who are not taking hormones.

Unfortunately, when I next saw Sally, she again complained of being depressed. "It's when I take my progesterone," she noted. "It's like I'm having PMS. I can't sleep and my energy is kaput." Complaints about progesterone are something I hear all the time. The miserable mood changes that some women experience while on progesterone can be a major complication of hormone replacement. In fact, I had prescribed micronized progesterone for Sally because it generally causes fewer unpleasant side effects than synthetic progestins. However, as Sally can attest, in some women even micronized progesterone can have negative effects.

One way to counter the side effects of these progestins is to carefully increase the estrogen dose. In a significant study, Dr. Barbara Sherwin demonstrated that the adverse mood changes in women receiving a progestin were less when their dose of estrogen was higher. With this in mind, I increased Sally's dose of Premarin. While this change definitely improved Sally's mood, she still felt weepy at times while taking her progesterone. While I'm certain that her ongoing divorce contributed to her unhappiness, Sally and I agreed that she was not functioning well and that progesterone was contributing to her troubles.

Another way to circumvent this problem is to administer progesterone in the form of a vaginal suppository. The theory behind this approach is that vaginal administration would concentrate progesterone's effect on the uterus, protecting it against cancerous change, while the blood levels reaching the brain would be less, resulting in fewer psychological and emotional side effects. When I switched Sally to vaginal progesterone, her moodiness largely disappeared. The last time I spoke with her, she was feeling vibrant, healthy, and had started exercising at a nearby gym.

Sally's case typifies the kind of thoughtful care required to provide patients with hormone modulation that is genuinely custom-tailored. It also typifies the kinds of issues that arise in finding the right preparations and combinations of estrogen, progesterone, and in some cases, testosterone. Doctors and patients must both be aware of the potential effects of any hormonal medication—both positive and negative—on mood states. This awareness should guide them as they develop a rational and effective form of treatment for the patient's suffering.

As I'll detail in chapters 8 and 9, individualized hormone modulation for mood disorder, or any other condition, involves attention to the type, dose, and schedule of progesterone. Alternatives to synthetic progestins are now available, including the oral micronized progesterone and vaginal suppositories of micronized progesterone. For some patients, like Sally, these preparations cause fewer psychological symptoms. Alternative schedules can also be tried. In some cases, progesterone can be dropped from the regimen, though such patients require very careful medical follow-up. The primary point is clear: Doctors and patients alike must be mindful of the mood-altering effects of progesterone, and make adjustments accordingly.

The Role of Testosterone

As a woman moves from perimenopause into menopause, estrogen is not the only sex hormone to drop precipitously. Levels of blood testosterone also fall, as much as 40 to 50 percent. When I evaluate a woman with mood disorder or sexual dysfunction, I not only measure her blood estrogen, I also measure her testosterone levels. Like estrogen, testosterone is produced in the ovaries and helps women maintain good energy and sexual function. Testosterone in women stimulates bone growth and muscle development, as it does in men. Testosterone also has an antidepressant effect on the female brain, producing a sense of well-being. Sufficient estrogen is necessary to stimulate receptors for testosterone and if her estrogen is low, a woman may not benefit from testosterone—she needs more estrogen to activate her testosterone receptors. The answer? A combination of estrogen and testosterone, and I can tell you from my clinical practice that it's a powerful therapy for many women with mood and sexual disorders. The analysis of twenty-six studies by researchers at Bowling Green State University, mentioned earlier in this chapter, proved that estrogen is an effective treatment for depression, but that combined estrogen/testosterone was even better.[17]

In her research, Dr. Barbara Sherwin has demonstrated the antidepressant effect of testosterone, and the estrogen/testosterone combination, in surgically menopausal women. I often treat menopausal women who are depressed or sexually disinterested, even while on hormone replacement. I am repeatedly impressed by the remarkable difference in their well-being once I add testosterone to their regimens. It clearly appears to have been the missing piece in their hormonal profiles, since their moods brighten and their sexual vitality returns, full force.

Julia was the first menopausal woman for whom I prescribed testosterone. When she came for consultation, she struck me as highly attractive and very bright, but she was despondent and seemed to lack spark. It was as though Julia had a vibrant personality trapped inside, unable to express itself. One month after I put her on testosterone, I saw the difference the moment she walked through my office door. She had the glow of vitality about her. "I can't tell you how much I've changed," she blurted out. "I feel so much better about myself. I'm bursting with energy and my husband is ecstatic because I'm interested in sex again." In chapter 4, I will further detail the exciting changes that are possible for women receiving treatment with testosterone. Also in chapter 4 is a complete discussion of testosterone's potential for men with mood, cognitive, and sexual disorders.

The Thyroid Connection

Two diseases of the thyroid can frequently be associated with psychiatric symptoms, including depression and anxiety. In *hyper*thyroidism, there is an overproduction of thyroid hormone, while in *hypo*thyroidism, too little hormone is produced.

Hyperthyroidism occurs most commonly in young adults as an autoimmune disorder called Graves' disease. Seventy-five percent of the cases occur in women. In older people, a toxic nodular goiter (irregularly enlarged thyroid) is the most common cause. Psychiatric manifestations of hyperthyroidism include an unstable mood, anxiety, irritability, restlessness, and hyperactivity. Some patients experience difficulty concentrating and are unable to focus. The symptoms in the elderly are quite different than in the young. They may appear depressed, apathetic, eat poorly, and lose too much weight.

Hypothyroidism is a deficiency of thyroid hormone, which can result in the slowing of most organs in the body. It is more common in the elderly and more prevalent among women. A number of conditions can produce this disorder, including surgical removal of the thyroid or radioactive iodine treatment of hyperthyroidism, resulting in destruction of a major portion of the gland. One form of hypothyroidism is Hashimoto's disease, an autoimmune disorder that produces a chronic inflammation of the gland. The disease is characterized by the presence of thyroid antibodies, which attack the gland, creating deficiencies of thyroid hormone levels.

The psychological symptoms of hypothyroidism may mimic depression; the person's mood is bleak and sad, and the patient complains of fatigue, weakness, and decreased sex drive.[18] In approximately 10 percent of patients, there may be paranoid delusions. Visual hallucinations and thoughts of suicide may hound the patient. Other clues that there may be a problem with a thyroid disorder include menstrual irregularities, worsening PMS, or a postpartum depression. These psychological symptoms often prompt doctors to refer women to either a psychiatrist or a therapist, but too often they never receive the medical evaluation that would turn up the underlying thyroid problem. If you have unexplained or difficult-to-treat mood disorders that fit any of these descriptions, including bipolar disorder, you should ask your physician or endocrinologist to examine your thyroid and run blood tests for thyroid hormone levels. The primary tests, which I will detail in chapter 8, are blood levels of thyroxine (T4), triiodothyronine (T3), and thyroid-stimulating hormone (TSH). In hyperthyroidism, T3 and T4 are elevated, and TSH is suppressed. In hypothyroid problems, T3 and T4 are down, while TSH levels are elevated.

The Role of Thyroid

In 1995, Haggerty and Prange commented that virtually 100 percent of patients with severe hypothyroidism also suffered from depression.[19] Psychiatrists and endocrinologists alike are gradually starting to recognize that thyroid problems, which lead to imbalances in thyroid hormones, can be the hidden cause of mood disorders.

The problem becomes apparent when we carefully evaluate depressed patients who don't respond to antidepressants. Among patients with treatment-resistant depressions who don't respond to any medications, we often find thyroid problems. Investigators have found that 13 percent of treatment-resistant patients have clinical hypothyroidism, as evidenced by elevated levels of thyroid-stimulating hormone (TSH). (Excess TSH is produced in a futile effort to stimulate more thyroid hormone.) In other words, their thyroids were underactive, producing too little thyroid hormone. (Some have a form of autoimmune thyroiditis, in which the thyroid malfunctions because it is attacked by the person's own immune system.) A smaller percentage of these treatment-resistant patients had *subclinical* hypothyroidism (slightly elevated TSH but normal blood thyroxine, the thyroid hormone), which may have been a factor in their mood disorder. Numerous studies have found that such patients benefit by receiving thyroid hormone along with their antidepressant.

A lack of thyroid hormone can lower the threshold for depression. On the other hand, either a deficiency (hypothyroidism) or an excess of thyroid (hyperthyroidism) may contribute to mania. A Dutch study of more than 3,700 psychiatric patients not only demonstrated a statistical relationship between low thyroid and mood disorder, it found that 18 percent of patients with low thyroid suffered from a severe form of manic depression, as compared to 0 percent of a control group. In another study of thirty patients with rapid cycling bipolar disorder, 41 percent were found to have some degree of hypothyroidism. This association could not be accounted for by treatment of these patients with lithium, which is frequently given to treat bipolar patients. Lithium in itself can induce hypothyroidism.[20] The use of thyroid hormone treatment has been reported to be successful in 50 to 65 percent of patients with rapid cycling bipolar disorder.[21] Why does thyroid matter to an individual's mental state? Thyroid hormones stimulate receptors for the neurotransmitter norepinephrine, and they appear to regulate serotonin receptors in parts of the brain that control emotions.

The message from these findings is clear: When depressed patients do not respond to antidepressant medications, they must undergo a complete medical evaluation that includes thyroid status. Any evidence of hypothyroidism must be treated with thyroid hormone, and the chances that their condition will finally respond to antidepressant treatment will improve.

ALLISON'S STORY: THE THYROID-BIPOLAR CONNECTION

When Allison was thirty-one she had her first episode of depression and was placed on amitriptyline, a tricyclic antidepressant. Within a few weeks she was having difficulty sleeping, her mind was racing, and she could not sit still long enough to accomplish anything. There clearly was something wrong that seemed to be made worse by her medication. Her psychiatrist told her that the amitriptyline had triggered a "manic episode" and that the treatment for this was a drug called lithium. Manic behavior is characterized by excessive physical and mental activity—constant walking, talking, writing—racing thoughts that jump from one subject to another without apparent connections.

Allison was a nurse and the idea of taking lithium frightened her. Lithium can have serious side effects—tremors, kidney damage, and abnormal thyroid function. But the doctor told her without treatment she would not recover. Though she resisted the idea that she was having a manic episode, deep down she suspected that the diagnosis was correct. She had always thought that her father had a bipolar syndrome, characterized by wide swings in mood from mania to depression, hence the term bipolar. Also she had identical twin brothers, one of whom was schizophrenic. Fortunately the other brother had always remained well.

Allison agreed to treatment and was started on lithium. The amitriptyline was discontinued and in its place another tricyclic antidepressant, imipramine, was prescribed. She gradually recovered from her manic episode but in subsequent years she suffered from recurring depression followed by bouts of manic behavior. Most frequently the depressive episodes occurred following ovulation. It seemed as though she had a severe form of PMS. At times during the premenstrual phase of her menstrual cycle she would become so agitated and angry that she would act out. On numerous occasions she would throw objects. One time she was sitting in front of her mirror and she hurled first her hairbrush and then a lamp at the mirror, shattering it. At other times she threw pieces of china. She told me with regret that she had destroyed several antique cups.

Because of the severe PMS she consulted a psychiatrist who was treating PMS with high doses of progesterone. She derived some benefit from the high-dose progesterone because the hormone produced sedation. She remained on this treatment for two years. Amazingly, in spite of her disability, Allison married and had two children, both daughters. During each pregnancy she felt well but in each case postpartum she developed a depression, which lasted four to five months.

At age forty-seven she consulted a new psychiatrist, who after listening to her story told her that she had bipolar disorder. He treated her with Depakote, a mood stabilizer that helped reduce her wide mood swings, and a combination of two antidepressants, Wellbutrin and Zoloft. There was some improvement in Allison's depression but she continued to have frequent episodes of premenstrual difficulty and on some occasions the depression could persist into other phases of the cycle. Usually, however, after she had her menstrual flow her moods improved for a short time. In spite of all her problems, Allison had worked continually full-time as a research nurse and had performed admirably raising her two young daughters. At times her depression became so severe that she had to fight constantly to avoid becoming immobilized. After one such episode when she failed to respond to treatment with Prozac and Paxil she was given electric shock therapy, which produced some transient relief.

After Allison had missed two menstrual periods and was having hot flashes and night sweats, her therapist suggested that she see me. The first time I met Allison in my office it was obvious that she was in terrible pain. Here was this clearly remarkably intelligent capable woman who was barely holding on. Her face told the whole story. Her constant pain was there for all to see. After I listened for a half hour to Allison's story of how she had struggled to keep working and be a good mother to her daughters in spite of all her devastating symptoms, I had to tell her, "You are remarkable." Some of the pain in her face gradually relaxed as we discussed the possibility that hormonal treatment might relieve some of her symptoms. Although after all her years of suffering, Allison was very skeptical about possible "cures." The blood testing clearly indicated that Allison was menopausal. She had an elevated FSH and a low blood estrogen level. Since thyroid problems can often be associated with menopause I was not too surprised to find that Allison's TSH was elevated. Also her blood thyroid levels were low, indicating an underactive thyroid. I was quite sure that both her menopausal status and her hypothyroidism were major contributors to her current condition. Both the hormonal states can markedly aggravate bipolar problems.

I started her on estrogen and thyroxine simultaneously. One month later when I saw her, the transformation was amazing. The painful facial appearance had been replaced by a large smile. "I can't believe this," she told me. "I feel transformed. Why hasn't anyone suggested this before?" That is a question I hear too often. All I can do is explain that many doctors have not yet associated hormonal problems with psychiatric difficulties. This is compounded by the exaggerated fear that estrogen therapy may place women at risk for cancer. Things are changing in this regard and I hope in the future I will hear fewer stories like Allison's.

I offer more information about the role of thyroid in hormone modulation later in this chapter.

HORMONE MODULATION FOR MOOD DISORDERS

Today, depression is a common household word. But we must remember that there is a distinct difference between the everyday use of the term (that is, "I'm so depressed!") and its meaning as a psychiatric diagnosis. In the world of psychiatry, serious depression is called major depressive disorder. What defines this disorder? A depressed mood, a decreased interest in pleasurable activities, or an inability to function that persists over a two-week period. This state of affairs must represent a genuine change in feelings and behavior. Furthermore, there must be at least four other symptoms present. Among them: weight loss, or less frequently, weight gain; insomnia; excessive sleep; fatigue; loss of energy; decreased concentration; agitation; lethargy; indecisiveness; feelings of worthlessness or guilt; and recurrent thoughts of death or suicide. These must impair the person's ability to work and/or have a normal social life. When these criteria are met, the person is said to have a major depressive disorder. When an individual has some of these symptoms but doesn't fulfill the criteria, he or she is said to have "depressive symptoms."

How do we determine whether mood changes are hormonal? First, I must emphasize that few family doctors, general practitioners, or psychiatrists consider the possibility that mood disorders have a hormonal component. This understanding has become the province of a few endocrinologists and psychiatrists working in the emerging field of psychoneuroendocrinology. That said, we believe that change is on the horizon. Until then, it's up to you to investigate whether hormones are a factor in your case. The only surefire way is to have your hormone levels tested, but you should consider certain criteria before pursuing such tests.

TABLE 1 *Hormonal and Nonhormonal Factors that Predispose to the Development of Depression*

Hormonal Factors	Nonhormonal Factors
1. Female gender	1. Family history of depression
2. Postpartum	2. Previous depressive symptoms
3. Premenstrual	3. Extended periods of stress
4. Perimenopausal	
5. Menopausal	
6. Hypothyroidism	
7. Hyperthyroidism	

In Table 1, I have listed the hormonal and nonhormonal factors that predispose to depression. As I've emphasized, sudden changes in estrogen levels, such as those that occur postpartum, prior to menstruation, and during menopause, can make women more vulnerable to depression. Certainly, if you think that you've experienced a decline in estrogen just prior to the onset of your depression, you have reason to suspect a hormonal factor. If you're a woman in your thirties or forties, and your depression is not responding to treatment, or it is accompanied by other unexplained physical or emotional symptoms, consider the possibility that you are perimenopausal, or perhaps experiencing an early menopause. (The latter is rare, but it's worth finding out about.) Either one of these phases is accompanied by drops in estrogen and/or testosterone that could contribute to your mood disorder.

When you are suspicious of a hormonal factor in your depression, it is crucial to obtain laboratory studies to bolster or disprove your clinical intuition. Table 2 lists the laboratory measures that I consider important in determining whether a patient's depression has a hormonal component. I suggest that you discuss these tests with your doctor and encourage him or her to consider ordering them. (More on this in chapter 8, which gives more details on these tests and provides reference ranges so you can evaluate your own results.)

As you consider whether your mood disorder involves hormones, see if you can define accompanying events—primarily, PMS, the postpartum period, perimenopause, or menopause. Here are factors to evaluate in each of these circumstances:

TABLE 2 *Blood Levels that Should Be Obtained to Evaluate for Possible Hormonal Components of Depression*

Assessment of Ovarian Function
1. FSH
2. LH
3. Estradiol (blood estrogen)
4. Progesterone
5. Testosterone
6. Free testosterone (See chapter 4.)
Assessment of Thyroid Function
1. Thyroxine
2. Free thyroxine
3. Triiodothyronine
4. Thyroid antibodies
Marker of Psychiatric Status
1. Platelet MAO

Premenstrual Tension Syndrome (PMS)

Since we know that declining estrogen can precipitate depressive symptoms, you can locate in Figure 4, which charts the menstrual cycle, the times when you are most vulnerable. The most sensitive time is the premenstrual days, followed by a short period right after ovulation. By far, the most common time during a woman's cycle for her to complain of anxiety and depression is the premenstrual period, labeled the Secretory Phase on the right side at the bottom of Figure 4.

If your estrogen levels are particularly low during this premenstrual phase, this strongly suggests a hormonal basis to your depression. Progesterone levels rise and fall during the two weeks before menses. In fact, it is likely that this rapid rise in progesterone can also trigger PMS-associated anxiety and depression.

I test for progesterone levels during the PMS period, and I evaluate the relationship between estradiol (the most common form of estrogen tested in blood) and progesterone. If a woman's estradiol levels are low

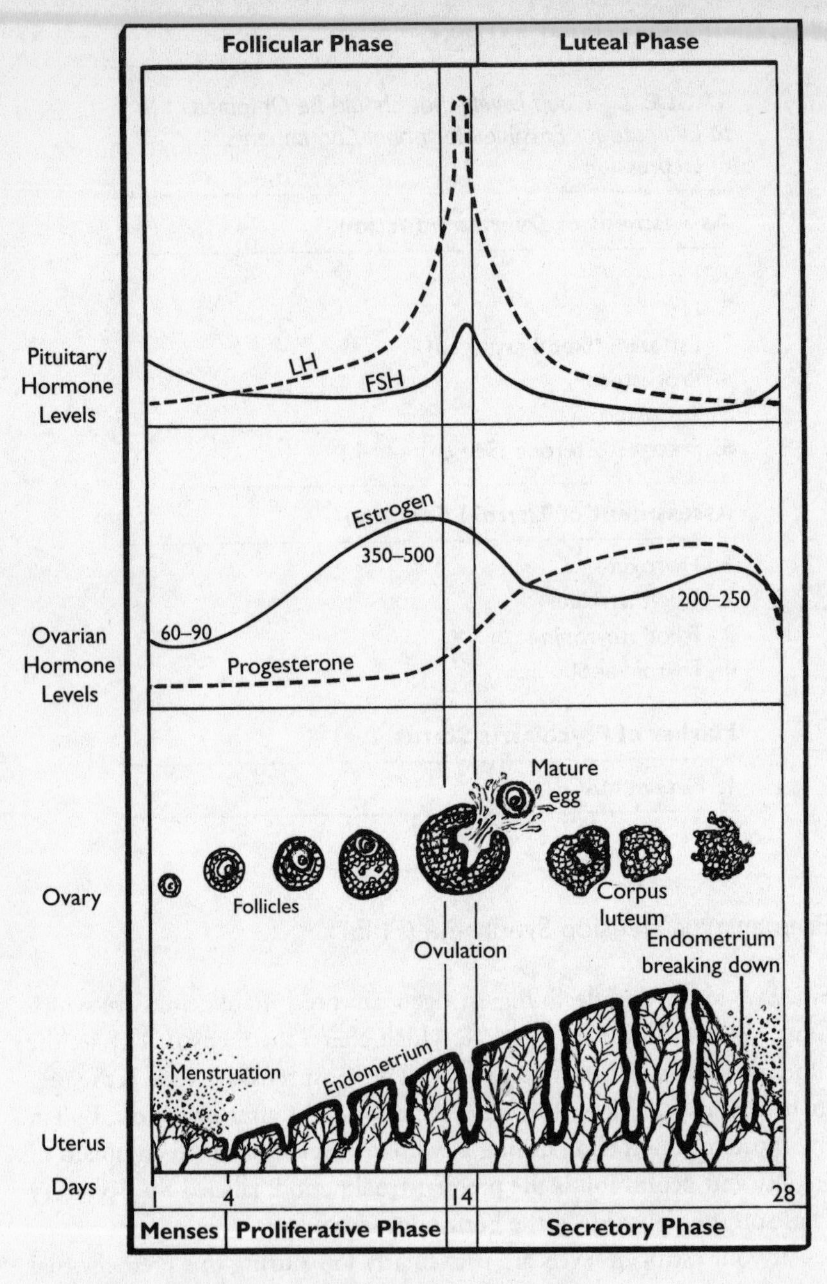

FIGURE 4 Cyclic Changes in Hormone Levels Across the Menstrual Cycle

Cyclic changes in pituitary (FSH and LH) and ovarian hormones (estrogen and progesterone), the ovary (ovulation), and uterus (growth in the uterine lining-endometrium) across a twenty-eight-day menstrual cycle. Peaks of FSH and LH midcycle produce ovulation following which the ovary secretes increasing amounts of progesterone during the premenstrual phase of the cycle.

during this time, then the depressant effects of progesterone will overshadow the antidepressive effects of estrogen. Try to have both your estrogen and progesterone levels tested, and find out whether, in fact, your estrogen is very low, your progesterone is very high, or both. This can help determine a rational course of hormone treatment. Having said that, it is not possible to identify which women will have PMS simply by examining their estrogen and progesterone levels. I will discuss in more detail the relationship between estrogen, progesterone, and PMS in chapter 5.

Perimenopause

If you are in your thirties or forties and unremittingly depressed, I suggest that you find out whether you are in early menopause or perimenopause. You can easily resolve this question with tests for levels of follicle stimulating hormone (FSH) and luteinizing hormone (LH), as well as estrogen. As women reach their perimenopausal years, the ovaries begin to shut down production of estrogen, and the pituitary gland secretes increasing levels of FSH and LH in a futile attempt to raise estrogen levels. Therefore, if you have high readings of FSH and LH, and low readings for estrogen, you are perimenopausal. (These tests will also reveal whether you've entered menopause. During actual menopause FSH and LH levels will be even higher.) An elevated level of the MAO enzyme is another clue that your depression involves hormones. If these findings are positive, consider the possibility that your hormonal dips are contributing to your mood disorder. You may also have your testosterone levels tested, since their decline during perimenopause can also contribute to mood disorder and sexual dysfunction.

Postpartum

If you are experiencing postpartum depression that does not remit after a few weeks, consider the possibility that your hormone levels have not returned to proper balance. Tests for estrogen, progesterone, and testosterone and thyroid should be ordered, though the estrogen test is most important. If estrogen remains abnormally low, then estrogen replacement, sometimes in conjunction with antidepressants, can be a highly effective treatment that resolves postpartum blues. Ellen's story illustrates the importance of hormone testing in postpartum depression.

ELLEN'S STORY: TREATMENT-RESISTANT POSTPARTUM DEPRESSION

Following delivery at the end of a pregnancy, a woman's estrogen levels drop drastically to levels that are one-hundredth of pregnancy levels, down to the values normally seen in premenopausal women. Depending on a host of predisposing factors, mind and body may not tolerate this radical change, and some women develop a postpartum depression. Ellen was such a woman.

The time I first saw Ellen, she was twenty-four years old, working as a nurse in a local hospital. Her periods had started at age eleven and were perfectly normal, but by fifteen, she was having mood-related symptoms of PMS—mainly depression and irritability. At twenty-one, she became pregnant, and she experienced well-being throughout the pregnancy. Right after giving birth, she had one of those brief periods of depression that is really quite common. The blues lasted several days, then seemed to disappear. However, within a short time frame Ellen's depression returned, and this time it would not go away. Ellen was extremely concerned about the persistence of her depression, since she had a family history of psychiatric disorders. As a child, her mother had such severe bouts of depression that Ellen was essentially raised by her grandmother. Ellen consulted her family doctor and was started on the SSRI antidepressant Zoloft.

Zoloft would not be the solution to Ellen's suffering. Several months after she had begun treatment, her depression worsened and she was in a constant state of fatigue. To add to Ellen's woes, she was diagnosed with hypothyroidism. Sudden drops in estrogen levels postpartum, and at the menopause, are frequently associated with the development of hypothyroidism. Her doctor prescribed thyroid hormone, but her depression still did not lift. When she finally arrived at my office, about six months after the delivery of her child, Ellen was in the throes of clinical depression. I ordered blood tests, which revealed that she had a very elevated platelet MAO level, a frequent finding in depression. Further testing revealed that Ellen's thyroid hormone dose was insufficient, and her estrogen levels were low. Since the thyroid gland controls the metabolic rate of all the other endocrine glands in the body, her continuing thyroid deficiency may have caused her ovaries to malfunction, so they were not secreting ample amounts of estrogen. The combination of a thyroid *and* an ovarian deficiency would almost surely induce depressive symptoms, particularly in a woman with a family history of psychiatric problems.

My first hormonal intervention was to increase Ellen's thyroid hormone dosage, followed by a prescription for oral estrogen. Since she'd complained of low energy, I switched her from Zoloft—a rather sedating antidepressant—to Prozac, a more stimulating one. After one month, her depression was already beginning to lift. Since she was taking estrogen, I also put her on oral micronized progesterone to protect her uterus. Within a few more months, Ellen's depression was completely resolved, and we were able to gradually taper her off estrogen and Prozac, while continuing her on thyroid hormone. This process was periodically guided—as it should be—by blood tests to make certain her hormone levels remained normal. I now see Ellen every six months, and I carefully monitor her thyroid function, which tends to vary. When her thyroid levels dip, she begins to experience mild symptoms of depression, but they are completely reversible with a simple increase in her thyroid medication.

Individualizing Hormone Modulation for Mood Disorder

A theme I will return to throughout this book is that every woman and man is a unique individual—in terms of genetics, biochemistry, hormones, and personality. We physicians must treat patients with this complexity and uniqueness in mind at every turn in the clinical endeavor.

What does this mean for hormone modulation? I tailor every patient's treatment to his or her unique biochemical, hormonal, and psychological profile. For patients with mood disorder, I must consider all the information from our one-on-one discussions, their physical examination, and their blood test results.

For example, thus far I've emphasized the special need for hormone modulation among women with treatment-resistant depression. These are patients who fail repeated trials with antidepressants, until we finally add hormones to the mix and they get well. But some treatment-resistant patients do extremely well with hormone modulation alone. In other words, their estrogen loss—or other hormonal imbalance—was the overriding cause of their mood disorder. All they needed was hormone therapy to correct the problem, and their depression lifted. These women never responded to antidepressants, and it turned out that they never needed them! What they really needed was hormone replacement.

But again, individualization is key, because many chronically depressed women need both hormones *and* antidepressants. If I determine that a particular patient has *no* hormonal disorder, then she clearly needs more creative or aggressive treatment with antidepressants—not hormones. Narrow-minded specialists always want to offer their one treatment for everyone's problem—the "if-all-you-have-is-a-hammer-everything-looks-like-a-nail" approach. As a clinician, I have tried to avoid this mistake, because I believe it could hurt my patients.

Normalizing Hormone Levels

Here is another principle I will repeat: I do not give estrogen, testosterone, or thyroid hormones to drive people to "maximal" blood levels. I don't believe I should be pushing patients to maintain superhigh hormone readings. Excessive hormones in the blood are likely to be as bad as insufficient hormones, and unsafe to boot. I base my determination of hormone deficiency on low blood readings for an individual patient, not on clinical intuition. And my goal is to restore her levels to within the normal range—not to exceedingly high levels. (In our study of severely sick patients with refractory depression, we used high-dose estrogen, but these special cases were exceptions to the rule, as I have explained.) When hormone modulation is carefully guided and followed with blood tests, the treatment has a rational scientific basis, which makes it safe and much more effective.

Moving Beyond Standard Doses of HRT

With this principle in mind, I don't treat women with mood disorders with the standard doses and preparations of hormone replacement therapy (HRT). I'll have much more to say about this in chapters 8 and 9, but the idea is not complicated. If I need to use higher-than-standard dosages of estrogen to bring a patient's estrogen level into the normal range, I will do so. If I need to prescribe an estrogen patch, or a different preparation of estrogen, I will do that, as well. For example, for some women, Estrace—an oral estradiol—is preferable to Premarin, the "conjugated estrogens" derived from equine (horse) urine, but for others Premarin is fine. I do not automatically rely on standard doses of HRT for women with serious depressive symptoms who require individualized therapy.

HORMONES FOR DEPRESSED MEN: WHAT WOMEN SHOULD KNOW

Hormones do not engage in gender discrimination. Men go through a form of menopause too, although the passage is much more gradual. The disparity in the rate and degree of hormonal decline accounts for the differences between the two sexes. In women, blood estrogen levels decline 70 to 80 percent within a few months following the onset of menopause.[22] In men, the blood testosterone decline is 40 to 50 percent and it occurs over years.[23] This may explain why relatively little attention has been paid to the "male menopause," whereas female menopause has become a sort of media darling. Partly because of their decline in testosterone, men at midlife often experience a variety of symptoms, including mood disorder, cognitive impairment, and sexual dysfunction.

Testosterone and Men's Moods

As they age, men are at greater risk of depression. How much of this increased risk is due to declining testosterone? With the current state of knowledge it is not possible to give a definitive answer. But some facts are known and they suggest a possible hormonal relationship of testosterone to depression particularly as men age. A study of 856 men aged fifty to eighty-nine found that older men had lower testosterone and higher rates of depression. How prevalent are low testosterone levels in older men? Twenty percent of men over age sixty have below-normal blood testosterone levels.[24] But this correlation is no guarantee that a causal relationship exists between low testosterone and a depressed mood. It is possible that the higher incidence of depression in elderly men is the result of factors other than low testosterone levels.

Overall, as I discussed earlier, the incidence of depression in men is less than women. We do know that men have 52 percent higher levels of serotonin in their brains than women, and research studies indicate that testosterone exerts a protective effect on the mood of men by enhancing neurotransmitter function.[25] Serotonin receptors in the brain are known to be enhanced by the presence of testosterone.[26] In animal studies, castration (which drastically reduces testosterone) increases brain levels of the MAO enzyme, which then inactivates the neurotransmitters serotonin and norepinephrine, which are needed for a balanced mood.

When scientists give these animals shots of testosterone, MAO levels return to normal.

I learned a number of years ago that testosterone's influence on mood was not limited to the aging male. After our successful demonstration that estrogen had antidepressant qualities in women, my colleagues and I embarked on a study to determine if testosterone had similar antidepressant qualities in men. We conducted a double-blind clinical study to compare the antidepressant effects of a synthetic testosterone, mesterolone, and a widely used antidepressant drug, amitriptyline. Thirty-four depressed men aged twenty-seven to sixty-two were randomly assigned to treatment with either mesterolone or amitriptyline. For the first two weeks, the men received a placebo in order to establish the baseline level of depression. For the next twelve weeks, patients received either the hormone or the antidepressant.[27]

Both treatment groups were equally depressed from the outset, but by the eighth week, both treatments reduced the men's depression significantly. This effect persisted throughout the rest of the study. Both the synthetic testosterone and the antidepressant were equally effective in healing the men's depression. However, there was a definite difference in the number of side effects reported. Only one mesterolone patient reported agitation while nine of the amitriptyline patients had a number of side effects, the most common being low and high blood pressure, rapid pulse, drowsiness, blurred vision, dizziness, impotency, and hypomania.

While we published this study in 1985, the use of testosterone-like drugs for antidepressant purposes in the United States is practically nonexistent. Mesterolone has been widely used in Europe, South America, and Japan for the treatment of depression in the aging male. Some of the reluctance on the part of psychiatrists and physicians in the U.S.A. can be explained by the turmoil and fear that has surrounded the use of anabolic testosterone-like steroids by athletes and bodybuilders.

There was also a concern that mesterolone use in young men might affect their fertility. In actuality, there was no significant suppression of sperm count with mesterolone. However, there was moderate suppression of blood testosterone and estradiol levels. Amitriptyline produced no significant changes in blood hormone levels and neither drug produced changes in blood cholesterol levels. Erectile disorder, one of the side effects observed with amitriptyline, is also seen with the use of many popular antidepressants like Prozac or Zoloft. None of the men treated with mesterolone had problems with erectile disorder.

What can we take from this study? The results suggest that a testosterone-like agent may be an alternative antidepressant medication for depressed males, who either don't respond to traditional antidepressants, or who suffer from disabling side effects from these medications.

Do Depressed Men Have Lower Testosterone?

Once I discovered that testosterone was an effective antidepressant, I believed it was necessary to find out whether depressed men had lower testosterone than nondepressed men. My colleagues and I compared blood testosterone levels in twenty-seven depressed and thirteen nondepressed men, whose mean age was approximately thirty-eight years.[28] We checked both their testosterone and estrogen levels, since estrogen can oppose the actions of testosterone. Our results were clear-cut: The depressed men had total and free testosterone levels that were 30 percent lower than those of nondepressed men, and their blood estrogen was 30 percent higher. Using sophisticated measurement techniques, we determined that the depressed men had low testosterone because they cleared the hormone more readily from the blood. We don't know the reason for this increased clearance. Free testosterone is the physiological active form of the hormone; the vast majority of testosterone is bound to a blood protein and is inactive. We do believe that high estrogen increases this protein binding, thus lowering the amount of available free testosterone. Theoretically, testosterone treatment increases the free testosterone that gets into the brain and stabilizes mood. Our clinical trial suggested that this is just what happens.

What to Do About Your Man's Depression

If your husband, partner, parent, or friend is depressed, and he either does not respond to antidepressants or has unacceptable side effects, he may need a hormonal evaluation. Find an open-minded endocrinologist, psychiatrist, or physician who will order hormone tests to find out whether he has demonstrably low total and free testosterone levels. (Low free testosterone is a particularly telling sign of hormonal imbalance.) If so, encourage the man you care about to pursue treatment with testosterone by discussing this option with his doctor(s). Use this book and the studies cited as a resource and motivator. No man should have to suffer endlessly with treatable symptoms of mood disorder, symptoms that make his life—and your life together—difficult to bear.

HORMONE EVALUATION FOR MOOD DISORDER: WHY WAIT?

Given what we are learning about hormones and mood disorder, I strongly urge doctors and patients not to wait long before considering a hormonal profile. With the advent of managed care, everyone wants to avoid unnecessary tests, especially if they are costly. But I have seen too many cases in which women and men suffered for years without proper hormonal evaluations that would have alerted them and their physicians to the underlying cause of their mood disorders. Countless trials of antidepressants, psychiatric visits, and lost work hours are costly, too. A battery of hormonal tests is a drop in the bucket compared with the emotional and financial expense of years of improper or inadequate treatment.

Given the current state of scientific investigation into hormones and the brain, and hormones and mood disorder, I could have devoted this entire book to the subject. The field is mushrooming, as we come to understand the intimate, intricate, and powerful influences of hormones on the brain. Hormones may be involved not only in depression, but in manic depression, mania, schizophrenia, and other psychoses. Dr. Deborah Sichel, a psychiatrist of the Hestia Institute in Massachusetts, has written an extremely helpful book called *Women's Moods: What Every Woman Must Know About Hormones, the Brain, and Emotional Health,* which describes the interrelationship between hormones and mental states in women.

As we continue to learn more, we must also take positive advantage of what we already know—that hormone modulation is a viable option for women with difficult-to-treat mood disorders who also have medically proven hormone imbalances. If you suspect that you are one of these women, become an informed and assertive patient who mobilizes your medical providers to take care of your needs. You can get the proper treatment, as long as you have the right information and procure the help of the right professionals.

3

Bolstering Brain Power

MEMORY AND MENTAL SHARPNESS

I realized long ago how much I learn by listening to my patients tell their stories. Given my specialty—the relationships between the mind, nervous system, and hormonal system—I've spent most of my clinical hours hearing women discuss the changes that occur in their lives at critical junctures—after giving birth; as midlife approaches; before and during the menopausal passage. From women undergoing "the change," I hear about the onset of hot flashes, drenching night sweats, and vaginal dryness, symptoms ranging from merely uncomfortable to intolerable. But I also hear about a whole set of symptoms that are not physical. I've already discussed the mood changes that can occur, but women also say that they can't recall things that happened days or even hours ago. They may enter a room for a reason and, once in the room, lose hold of that reason. They suffer from an inability to focus, sometimes called "muddled thinking." They may have trouble making simple, everyday decisions, as well as the complex, major decisions about work, family, or health. They find themselves committing verbal slips and uttering malapropisms. All too often, these cognitive difficulties are more than passing symptoms, like the occasional stomachache or skin rash. They become serious, even debilitating conditions that affect women's ability to work and enjoy their lives.

When listening to these women, two thoughts come to mind. First, their symptoms are clearly linked to the loss of estrogen. Among these women, who are mostly in their thirties, forties, and fifties, blood tests usually confirm that their ovaries are in the process of shutting down, so their estrogen is low or fluctuating. Second, I recognize that these cognitive symptoms, which often occur in tandem with depression, are damaging to a woman's identity. The physical symptoms can be debilitating, but the mental ones can be demoralizing. A woman's sense of self is bound to her ability to remember, to make decisions, to think and speak clearly. For women whose menopausal transition is a time of cognitive impairment, the experience is confusing, frightening, and immensely frustrating.

A woman in this circumstance not only feels that she's lost certain capacities—she feels that she's lost a major part of her identity. It makes matters worse when her doctor doesn't recognize or validate her symptoms. Until recently, this was both understandable and forgivable. Physicians, psychiatrists, endocrinologists, and gynecologists knew little about research on estrogen and the brain, and they could not be expected to view women's cognitive complaints as a result of hormonal shifts. Thus, doctors in many specialties tended to ignore and misdiagnose women's cognitive symptoms. The problem is, today's doctors often continue in this vein, even though research on hormones and the mind has advanced in leaps and bounds. (I'm not laying blame; it's largely a matter of the lag time between breakthroughs in research and changes in clinical practice.) Good gynecologists certainly know that "muddled thinking" can bedevil women at menopause, but few realize that this problem can arise many years before frank menopause, and rarely do they implement a careful regimen of hormone modulation for these patients. As a result, women who experience a loss of memory and mental sharpness don't get the understanding or the medical treatment they need and deserve.

The cognitive effects of hormone imbalances—mainly low and/or fluctuating estrogen, but problems with other hormones, as well—are both varied and profound. What are the symptoms? Here is a brief list:

- *Forgetting things you said or did within the past few days, hours, minutes.*
- *Losing the train of a conversation, so you have to ask people to repeat things they've just said.*
- *Misplacing things within moments of putting them somewhere, such as keys or eyeglasses.*
- *Forgetting names, sometimes even your own family members'.*
- *Finding that memories don't come to you as accurately or quickly.*
- *The "it's on the tip of my tongue" experience of being unable to get words or thoughts out.*

- *Saying malapropisms—wrong words that sound similar to the ones you intended.*
- *Difficulty focusing or concentrating on work or other tasks.*
- *Feeling hazy or fuzzy-headed, so that your thought processes seem foggy.*
- *General absentmindedness.*
- *A change in faculties such as reading ability, driving, or writing.*

The women in my practice describe their symptoms, which often include one or many of the above. I order blood tests that usually confirm a hormonal imbalance, mainly low estrogen, and commence with a treatment regimen specifically designed to correct that imbalance. In a matter of weeks or months, these women generally experience noticeable or even remarkable remissions of their cognitive symptoms. I don't claim to "cure" every patient of all her cognitive impairments, but hormone modulation is clearly the best available treatment when these symptoms can be directly traced to hormone imbalances.

HORMONES FOR COGNITION: EARLY CLUES AND EXPERIMENTS

"Doctor, I think I'm losing my mind!" I still remember the pained expression of the first female patient who shared this fear with me. Leila was a fifty-two-year-old businesswoman who had undergone menopause two years earlier. She was a highly intelligent, self-assured woman, president of her own company, an executive dealing with many complex problems every day. She told me a story that typified her troubles.

"My memory has gone," she said. "The other morning my husband was in the bathroom shaving when I went in to take my shower. We started telling each other our plans for the day, as I turned on the water. We were chatting while I stepped into the shower, and he stopped me in mid-sentence: 'What are you doing?' 'I'm taking a shower, what do you think I'm doing?' 'But you just took a shower three minutes ago.' I couldn't believe it. I had taken my shower, went out to get dressed, then totally forgotten what I had just done. For the first time, I thought to myself, 'I had better do something about this.' I have too much responsibility at work to have these kinds of memory lapses." Fortunately, Leila's story had a happy conclusion. I prescribed estrogen treatment for her, and in a matter of weeks, her memory was literally restored. Leila returned to her old habit of showering only once each morning.

By that time, I had conducted sufficient research in hormones and the brain to believe that hormone replacement could potentially reverse the kind of cognitive symptoms that afflicted Leila. Don Broverman, Bill Vogel, and I had conducted our studies of automatization in male subjects, which demonstrated that men with higher testosterone were more adept at a particular set of cognitive tasks—namely, those that depend on automatic behaviors such as walking, talking, reading, writing, and maintaining one's balance. We took this a step farther, giving infusions of testosterone to young men and demonstrating that they performed better on tests of "automatization" than men who got a placebo (saline) infusion. We also showed that testosterone levels had an influence on certain brain wave patterns (as measured by electroencephalogram [EEG]) that were characteristic of people with particular cognitive abilities, like automatization. It was becoming powerfully clear that hormones had a significant role in the brain, and we expanded our research to study the effects of estrogen on women's brains.

My colleagues and I would take advantage of the fact that women's estrogen and progesterone levels naturally rise and fall, in predictable patterns, during their monthly cycles. (See Figure 4, page 60.) We decided to evaluate the thinking processes of eighty-seven women during the estrogen and progesterone peaks in their cycles.

Among ovulating women, estrogen peaks at mid-cycle—just before ovulation. We theorized that women would do better on automatic tasks during this estrogen peak. Based on our prior research, we also predicted that they'd do worse on a different set of tasks, known as "perceptual-restructuring" (or PR), which required them to set aside automatic responses to obvious stimuli in favor of more complex responses to less obvious stimuli. (You may best understand the difference between these two mental abilities by virtue of the tests we used. Good "automatizers" name the colors on flash cards more quickly. Women good at PR are better able to identify figures hidden within a set of pictures—they can detect more subtle visual impressions.) Since progesterone *blocks* the action of estrogen, we predicted that when progesterone peaks during the cycle (about a week after ovulation), we would find the opposite: Women would be *worse* at automatic tasks and *better* at PR.

We confirmed that women did do better on automatic tasks, and less well on perceptual-restructuring tasks, when their estrogen levels peaked around day ten of their cycles. When their progesterone levels were cresting, at day twenty, the opposite was true—their PR ability was much better. (Interestingly, we did not find the predicted effects on days that were too far away from the peak levels of estrogen and progesterone. Also, we did not see these effects in women who were not ovulating, and, therefore, not undergoing the usual rise and fall of estrogen and progesterone.)

What did this research mean? Put simply, some cognitive functions cycle with a woman's hormones. Just as estrogen and progesterone have powerful effects on certain "target" tissues in women—the uterus, breast, heart, and bones—they have powerful effects on their brains. Our research also showed that an "estrogen good, progesterone bad" interpretation is a bit too simple. First, automatization and PR are both useful and necessary mental abilities. One is not "better" than the other. Second, each hormone has a role to play in mental functioning across the monthly menstrual cycle. The trouble occurs, we have found, when the rise and fall in estrogen and progesterone premenstrually trigger PMS or when estrogen is too low *throughout* the cycle, as occurs during perimenopause and menopause.

At the beginning of our research, I began to treat women with cognitive impairments—the loss of memory and mental sharpness—with estrogen and other hormonal medications. My clinical treatments were never designed on the basis of theory or guesswork—always on careful readings of hormone levels. I prescribed medicines designed to rectify real deficits in these women's hormone levels, whether they had low estrogen, testosterone, thyroid, or other endocrine imbalances. While we did not yet have a sizable body of research proving that hormones restored memory and mental sharpness, the treatments were justified as forms of hormone replacement for women with a host of symptoms associated with the menopausal transition. But I often observed the cognitive changes that we had predicted—clear improvements in short-term memory, mental focus, and decision-making. Later, we would have the clinical proof we needed that such treatments were effective for cognitive problems.

SHEILA'S STORY

Consider just one of my cases, Sheila, a fifty-year-old businesswoman who was in marriage counseling with Bill Vogel. Bill thought she needed to have her menopausal symptoms evaluated by me. A year earlier, she began having severe hot flashes and night sweats, which interrupted her sleep seven to eight times a night. Her gynecologist had started her on low-dose estrogen, and cycled her with a synthetic progestin for twelve days each month. In spite of her estrogen treatment, Sheila continued to endure hot flashes and night sweats, which became more frequent when she was taking the progestin. To get a good night's rest, Sheila had begun taking a sleeping pill. During this difficult period, she was also having marital problems, and her physical and emotional stress took a toll. When she became overtly depressed, her physician prescribed Prozac, but she soon stopped taking the antidepressant because it was having a sedative effect during the day.

One other symptom added horribly to Sheila's stress: her increasing lapses of memory. Sheila's memory problems had begun on the same day she experienced her first hot flash—no bizarre coincidence. In an incident of forgetfulness she will always remember, friends had asked her to baby-sit their cat for a weekend. "This was something I looked forward to doing," she said. "Yet I absolutely forgot to feed and water that cat. I was devastated."

The cat survived, but Sheila was not so sure about the long-term health of her memory. For twenty years, she had run small businesses, tracking hundreds of details without missing a beat. But by age fifty, those details began to elude her. An accumulation of small lapses convinced her that the cat incident was no aberration. Her concerns grew with each slip—such as the time she drove to the bank without her deposit, or continually forgot about a tenant's request for a repair. "It was maddening, especially because I'd always been so sharp," said Sheila, who co-owns sixty apartment units. In addition, her husband had involved her in the management of a restaurant he had purchased, and the restaurant was losing money. "Normally I would have been able to adjust to these problems, but I can't do that now. I'm totally losing it and I'm afraid that shortly I'll be gaga."

Sheila's case exemplifies the problem women face as they seek appropriate treatment today. Unfortunately, many doctors, regardless of specialty, would hear that Sheila was already taking HRT (estrogen and cyclic progestins), and assume that hormones could not be a factor in her problems with mood, memory, and mental sharpness. But I tested her serum estrogen levels and found them to be quite low! The problem was not complicated: The dose of estrogen prescribed by her gynecologist was simply inadequate. It had not raised her estrogen levels sufficiently.

Over three decades of clinical experience, I have learned to honor a basic medical truth: Every patient is unique, and the right dose of medicine for one individual won't necessarily be right for another. I increased Sheila's estrogen dose in two stages, and it wasn't until she was taking the higher dose that her hot flashes and night sweats finally stopped. At the same time, this higher dosage had raised her blood estrogen levels *into the normal range.* Again, my goal was not to push Sheila to superhigh estrogen levels, rather to *normalize* these levels. Not only did Sheila's physical symptoms abate, her short-term memory returned to normal! She began to function as she always had in the past: with great energy, efficiency, and mental acumen.

ESTROGEN AND THE MECHANISMS OF MEMORY AND COGNITION

To appreciate estrogen's effect on memory and mental sharpness, it helps to first understand how the brain stores memory. The primary memory storage centers are located in the brain structures called the hippocampus, mammillary bodies, limbic system, thalamus, and amygdala. (See Figure 5.) Neuroscientists have identified these memory "centers" from studies of people who've suffered impaired memory due to brain damage or injuries following a stroke, infection, lack of oxygen, surgery, and Alzheimer's disease.

What, exactly, does estrogen do in the brain to influence memory and thinking? Consider the following roles played by this sex hormone:

- Estrogen increases a chemical enzyme (known as "choline acetyltransferase") needed to synthesize the neurotransmitter *acetylcholine.* Acetylcholine is *the* neurotransmitter most critically involved in our memory functions. In patients with Alzheimer's disease, whose major symptom is memory loss, acetylcholine is tragically deficient. Animal studies have proven that estrogen increases the number of acetylcholine-producing nerve cells in areas of the brain critical to memory, including the hippocampus.
- Estrogen also enhances the availability of acetylcholine by ensuring that brain cells take up choline, which is essential for the synthesis of acetylcholine. In animal studies, estrogen deficiency reduced the

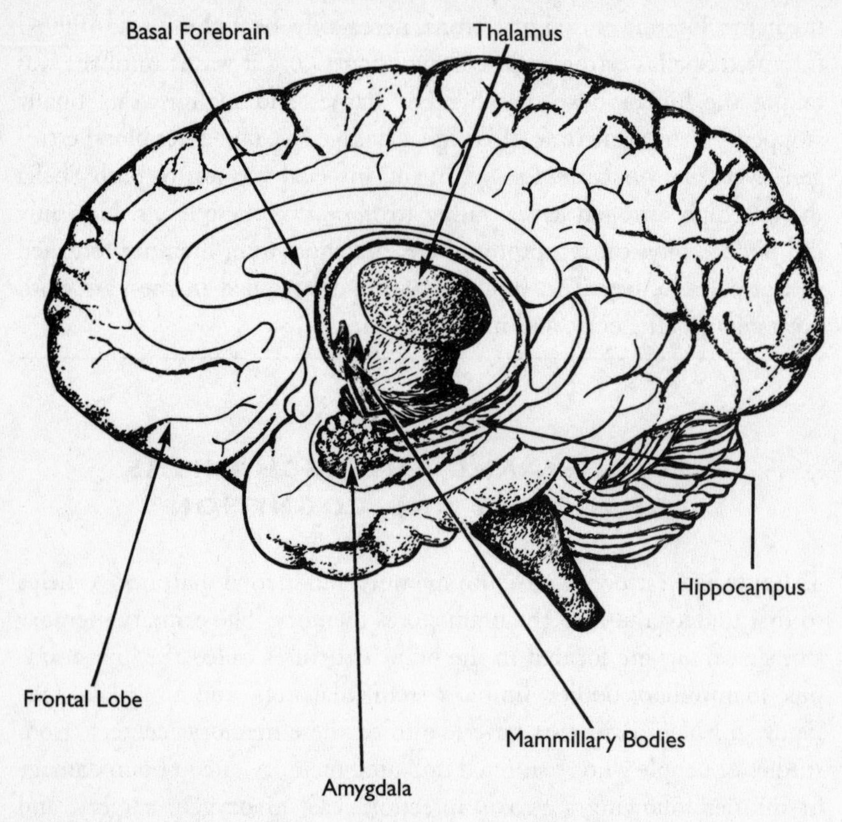

FIGURE 5 Areas of the Brain That Control Memory

uptake of choline by cells in the hippocampus. When estrogen was given to these same animals, this biochemical problem was corrected—which resulted in more acetylcholine. Generally speaking, more acetylcholine means better memory.

- Estrogen facilitates the networking between nerve cells. How so? It enhances the growth of dendritic spines on neurons, finger-like projections that form synaptic connections with other nerve cells. (Both the long appendage called the axon and the smaller outgrowths called dendrites make synaptic connections.) More synapses mean more docking sites for transmitting chemical "information" from neuron to neuron, essential for sound cognition and memory. (See Figure 1, page 16.)

- Estrogen promotes the actions of several "nerve growth factors," which protect against the damaging effects of age and injury on "cholinergic brain cells"—ones that secrete and receive acetylcholine. These cells are critically involved in the transmission and storage of memories.
- By inhibiting the enzyme MAO, estrogen enhances the availability of the neurotransmitters serotonin, norepinephrine, and dopamine, which not only modulate mood, but which are essential in the mapping of interconnecting neural pathways involved in memory.
- Estrogens act as natural antioxidants that protect nerve cells from free radical damage, which has been implicated in memory loss associated with Alzheimer's disease and aging.

And this is not even a complete inventory! Estrogen is to the brain what nutrients are to the whole body—fuel for so many basic functions.

Recent advances in high-tech diagnostics now permit neuroscientists to "look" at brain activity without having to surgically open the skull. Using MRI (magnetic resonance imaging) techniques Sally Shaywitz and her colleagues at Yale University were able to assess brain activity in women performing memory tasks, involving the recall of verbal material. As mentioned in chapter 1, Shaywitz studied a group of postmenopausal women who had not taken estrogen for months, then gave them estrogen (Premarin) for a period of three weeks, followed soon thereafter by a three-week period of "treatment" with a placebo. The researchers found that women on estrogen—as compared to the same women on placebo—showed a marked increase in brain activation, visible on MRI. The Premarin dosage they used—1.25 mg daily—is similar to the dosages of estrogen I have frequently found to be effective in treating my patients with memory loss, cognitive dysfunction, and mood disorders.

However, the fact that scientists have proved that estrogen is involved in the basic brain mechanisms of memory and cognition does not, ipso facto, mean that *giving* someone estrogen will reverse problems of memory and cognition. Scientists need to demonstrate that, first in animal studies and then in human clinical trials.

Estrogen has been available for human treatment for over fifty years. Even back in the 1940s, studies of elderly women suggested that estrogen might have a significant impact on the brain. Estrogen therapy caused a distinct improvement in memory among women living in a home for the aged, while those who did not receive estrogen experienced a marked deterioration in their memory.[1] Among the treated women, when estrogen was withdrawn and testing was repeated a year later, their memory capacity dramatically deteriorated—to levels even lower than the original baseline measurements two years earlier. How ironic that estrogen's effect on the brain was apparent fifty years ago, yet today many skeptics question the beneficial effects of estrogen on the central nervous system—even in the face of accumulating modern-day evidence.

There was a long lull in research on hormones and the brain between the 1940s and 1970s. During this lull, research on the damaging effects of aging on memory almost universally failed to examine the role of the brain's hormonal milieu. A number of double-blind studies starting in the mid-1970s reported a significant improvement in memory among menopausal women treated with estrogen. One of these studies found that estrogen was superior to a placebo in improving the memory capacity of postmenopausal women.[2] Yet a number of studies during this time failed to find any positive effect of estrogen on memory. This failure may have been a result of the type of tests used, which assessed aspects of memory that we now believe are *not* influenced by hormones. We have a more sophisticated understanding of memory, and it has become crystal clear that only certain aspects of the multifaceted memory process are regulated by estrogen.

In the past two decades, well-conducted research studies have given us strong evidence supporting hormone treatments for memory in select groups of women. I venture to say that most doctors and their female patients are not fully aware of the scope and quality of the human research on estrogen and memory. In the following section I offer a brief overview of these studies, one that I believe offers hope for countless women confronted with these distressing symptoms that hamper their quality of life.

Estrogen and Memory: The Evidence

MELANIE'S STORY

Melanie walked into the room, and then stopped suddenly. Why did I come in here? she wondered. It was the fifth time she'd repeated this scene in the previous two weeks. She looked around, desperately trying to find an answer to her question, but nothing triggered her memory. Seconds before she knew why she'd entered the dining room, but now it was gone. Just as suddenly, it came back: She needed a plate for hors d'oeuvres from the hutch. Melanie was grateful, for in recent times there'd been instances where the reason she walked into a room was lost, never to return.

Melanie's memory was a victim of hormone depletion. Six months earlier, her periods had become irregular; in the prior three months, they'd stopped completely. Her face would become warm and flushed, and she often woke in the middle of the night drenched in a cold sweat. Her body's thermostat was fluctuating wildly, and she was either too hot or too cold. She was not only forgetting why she entered rooms but where she put things away. All sorts of household items seemed to disappear, and there was no burglar in sight.

Melanie's relatives were coming for Christmas dinner. The previous summer at the family reunion she took a slew of photographs they had never seen. She remembered putting them away, but on the eve of her relatives' arrival, she didn't have a clue as to where. It was far from an isolated incident, and Melanie was becoming frightened. The memory problem was even affecting her work as a clinical social worker. She began to agonize about her job, feeling that memory loss and inability to concentrate were causing her to let her clients down. In the midst of a session, she'd forget some critical piece of information just imparted by a client. It had become so bad that she began to see fewer clients, until the size of her practice dwindled, along with her self-esteem. Melanie had always taken great pride in her ability to help people in difficult straits. As she began to search for solutions, she read that one of the changes associated with menopause is memory loss. With this realization, Melanie eventually found her way to my office.

Estrogen's positive effects on memory would be the solution to Melanie's memory problems. As I've described throughout this chapter, estrogen has both a biochemical and anatomical influence on the memory process. When she came to see me, Melanie's blood estrogen level was quite low, and her blood level of the pituitary hormone FSH was definitely elevated. These critical laboratory findings unequivocally establish the fact of menopause. Melanie was also having rapid mood swings and a declining sex drive—typical symptoms during "the change." Melanie would require a fairly high dose of estrogen to correct her memory problem, but we have learned that the brain often requires larger doses of the hormone than other organ systems of the body. I can frequently reduce this higher dosage once the desired therapeutic effect has been achieved, and that was the case with Melanie. After several months on estrogen, Melanie stopped finding herself in rooms for no reason, and her mental concentration, which enabled her to work at full capacity, had completely returned. Her physical symptoms were relieved and her sex drive was revived to boot.

Melanie is but one of scores of my patients with similar stories. Perhaps you face similar difficulties. Do you find yourself forgetting where you left your keys or glasses? Are you hounded by an inability to recall names, places, or recent events? Do you constantly turn to your mate or close friend in social circumstances with such questions as, "What's that person's name again?" "When did we go on that vacation?" "Where in God's name did I put my scarf?" or "When was I supposed to pick up the kids?" Do you make errors at work because you forget significant details? Have you made a list in order to remember things to do, then forgotten where you put the list? Have you ever been embarrassed in a work situation because you forgot something obvious, or simply couldn't find the right word, in the midst of a meeting or conversation? Depending on your age and hormonal status, a decline in gonadal hormones—mainly estrogen—may be your problem, and the key to your solution.

Case histories such as Melanie's lend credence to the idea that hormone modulation can resolve memory loss for many women. But I am a clinician/scientist, and while the clinician in me finds these case histories very compelling, the scientist in me wants more hard-core evidence. This evidence usually begins with individual research studies and culminates in clinical trials that prove the efficacy of a hormone such as estrogen for a specific indication, such as memory loss. A growing body of data supports the memory-enhancing benefits of estrogen for many women. It is time for clinical trials that prove estrogen's positive effect on memory.

Dr. Barbara Sherwin's Studies on Verbal Memory

Research studies of estrogen's effect on memory have been spearheaded by the farsighted experimental psychologist, Barbara Sherwin, Ph.D., of McGill University in Montreal. Dr. Sherwin has evaluated different types of memory—including "verbal memory," which involves recall of words or the content of short paragraphs, and "visual (nonverbal) memory," which involves the recall of visual impressions. She has tested different populations of women undergoing estrogen decline—mainly those experiencing the gradual loss that occurs with natural menopause, and those who experience the sudden loss that occurs after surgical menopause (a complete hysterectomy with removal of the ovaries).

During the natural menopausal transition, ovarian production of estrogen shuts down over many months, and even after menopause, the ovary still produces small amounts of estrogen for years. When the ovaries are surgically removed, however, hormone levels drop precipitously; only a small amount of estrogen remains, that supplied by the adrenal glands. Thus, surgical menopause is a sudden and drastic passage, as opposed to the gradual passage of natural menopause.

In a study on women undergoing surgical menopause Dr. Sherwin's findings were particularly compelling. Surgical menopause, therefore, presented Dr. Sherwin with a perfect research opportunity: She would administer tests to women just before they underwent total hysterectomy with removal of the ovaries, then redo the tests months afterward. This would enable her to accurately gauge the effects of rapid estrogen decline on memory and cognition. She not only determined whether the loss of ovaries caused memory problems, she could compare estrogen users to nonusers, to find out if estrogen protected these women against any loss in memory functions.

In three studies of women mostly in their mid-fifties who were undergoing surgical menopause, Dr. Sherwin found surprisingly consistent results: The loss of estrogen caused women to suffer deficits in their verbal memory, but this memory impairment was prevented by treatment with estrogen.[3] In two of these studies, women taking estrogen not only suffered *no* reversal in their verbal memory after surgery, they actually had *better* memory than before surgery!

Dr. Sherwin made a unique contribution to the understanding of how hormones affect memory when she undertook a controlled study of fifty women undergoing surgical menopause who received shots of either estrogen, testosterone, a combination of estrogen and testosterone, or a placebo.[4] (Remember, women whose ovaries cease to function not only lose estrogen, they also lose testosterone.) Months after surgery, women taking any of the hormone treatments *experienced no setbacks whatsoever* on two tests of short-term verbal memory, one test of long-term memory, and a test of logical reasoning. By contrast, women in the placebo control group—who got no active hormone—experienced setbacks in all four of these tests.

One interesting and important side note to this study: Dr. Sherwin tested hormone levels in all the patients, and those who got hormone shots after surgery were found to have normal levels of the hormones they received—estrogen, testosterone, or both. (Of course, those who got only a placebo after their ovaries were removed had extremely low levels of these sex hormones.) This proved that the women did *not* need super-high hormone levels to retain their normal memory capacities—they needed only *normal* hormone levels (what doctors call *physiological* levels). As a clinician treating women for mood and memory disorders, that has been my guiding principle.

Research on Visual Memory

Dr. Sherwin has not found compelling evidence that hormone therapy protects against a decline in *visual* (as opposed to verbal) memory. However, a recent study led by Susan Resnick, M.D., and her colleagues at the federal government's National Institute on Aging (NIA), suggests that estrogen replacement may indeed protect visual memory, at least to some extent.[5]

The research by Sherwin and Resnick is far from the only data supporting hormone modulation for memory loss. There are at least fifteen studies, of different types and designs, that have considered whether estrogen-based hormone treatment improves memory and/or cognition, and 80 percent of them have shown positive results on at least one test. The studies are not uniformly perfect in their methods or stunning in their results, so there is no doubt that much more research must be done. We certainly need more refined information on the best hormone regimens to prevent or treat memory loss. But in the meantime, I believe that we currently have sufficient evidence to support the use of hormone modulation for women who have clear-cut signs of hormone-related decline in memory.

WHAT TO DO ABOUT MEMORY LOSS

Certainly, not all short- or long-term memory loss has a hormonal cause and cure. But it is my contention that a much-larger-than-appreciated percentage of women with memory loss *do* have a hormonal problem, and in most cases *it can be successfully treated.*

How can you tell whether your memory problems have a hormonal cause? The only surefire way is to have your hormone levels tested—low and/or fluctuating levels of estrogen, and perhaps also testosterone, are strong indicators. But you and your doctor might not consider hormonal blood tests unless you have reasons to suspect a problem, and this is understandable. If you are experiencing memory lapses, ask yourself these questions:

- Are my memory lapses frequent and alarming, or are they only occasional and not very severe? (In other words, do I have a real memory problem or just sporadic moments of forgetfulness that are no different from before?)
- Do my memory problems seem to occur during specific times in my cycle, such as the premenstrual days?
- Am I definitely approaching, or well into, menopause?
- Am I in my forties or early fifties, not sure whether I am perimenopausal, but suspicious that I am, based on various other symptoms (i.e., hot flashes, night sweats, vaginal dryness, sexual disinterest or dysfunction, or mood swings)?
- Am I in my thirties, yet experiencing symptoms that I know might be perimenopausal? (Early menopause is rare but it does happen.)
- Have I been having these memory lapses in the months after giving birth?

An affirmative answer to any one of these questions is a strong clue that you may be experiencing a decline in sex hormones; a decline may be one cause of your memory problems, if not *the only* cause.

The next step, then, is to discuss these concerns and suspicions fully with your primary care physician and at least one specialist, either your gynecologist, endocrinologist, or psychiatrist. In short, your goal is to find one or two open-minded physicians who are willing to consider the hormone connection in your case, who will then order the necessary hormone tests, including blood levels of estrogen and testosterone, as well as thyroid hormones.

Once you have your blood test results, you can know with certainty whether your hormone levels are low. If the results are equivocal, it makes sense to have the tests repeated, since fluctuations in estrogen and testosterone are common, not only within the menstrual cycle. For instance, perimenopausal women may need several tests, since their hormone levels may vary month to month—sometimes wildly. A fluctuation in estrogen levels and an increase in FSH levels are themselves a sign of perimenopause, and an indicator that hormones may be involved in your memory problems.

Clear-cut evidence that your hormones are low, out-of-balance, or fluctuating is your strongest piece of evidence that your memory problems are hormone-related. The next step is to work with your health care providers to get hormone modulation geared specifically to normalizing your estrogen levels, and if need be, your testosterone and thyroid levels, as well.

Day to day, week to week, year to year, my clinical experience convinces me that this approach is a rational, scientifically grounded way to treat hormone-related memory loss. I cannot claim to "cure" every patient of all her memory woes, but I have had consistently good results with a high percentage of the women who come to me with these complaints. Nor is the treatment always clear-cut; often, I must work closely with each woman to arrive at the right dose of the right hormone medications—the regimen that restores her memory functions and well-being.

Unfortunately, doctors and patients alike are not well informed about the causes and proper treatments for memory loss. Memory problems are, to some extent, a normal part of aging. But women at midlife should not have to put up with serious memory problems! Doctors should not tell such women that their memory loss is "only in their head" or "nothing serious," when most often their problem is a *treatable hormone imbalance.* Given the incidence of women's memory problems, it's no wonder that the health and women's magazines keep trumpeting new "natural cures" for the widespread problem of memory loss. A few of these treatments, including the herb ginkgo biloba, have some studies that support their efficacy, but for most of these "cures" the evidence is weak.

In my view, when hormone loss is the real cause of a woman's memory loss, natural treatments that don't rectify the hormonal problem won't be effective enough. It's like trying to treat iron-deficiency anemia with vitamin C—it may help you feel a bit better, but it won't reverse your iron deficiency. What you really need is iron. Likewise, when your memory problem is caused by estrogen loss, what you need is estrogen.

LYNN'S STORY

Misunderstandings about the true causes and best treatments for memory loss can cause confusion and suffering. One of my patients, Lynn, sixty-one, was so upset by her memory loss that she believed she was developing Alzheimer's disease. Lynn first became aware of a memory problem in her early fifties, and one particular incident had stuck in her mind. It was

Thanksgiving, and she and her husband and their two kids were leaving on a trip to visit Lynn's parents. They planned to depart immediately after her husband came home from work, and Lynn was slated to pick up her son at soccer practice so there would be no delay. When she heard her husband's car coming up the driveway, it suddenly struck her—she'd completely forgotten to pick up her son. This was but one of a number of incidents that convinced Lynn that her memory was in trouble. By the time I saw her, Lynn's ability to remember had further declined. "People tell me things and ten minutes later I've totally forgotten what they said," she told me. "My children are constantly saying, 'Mom, I just told you that!' " Lynn was getting really worried about Alzheimer's.

I took a careful medical history from Lynn. Her periods had stopped at age seventeen, and she did not menstruate again until twenty-five. (It is not entirely uncommon for teenage girls to stop having their periods when they go away to college; the stress of change may be the cause.) After she married, Lynn tried to become pregnant but had two miscarriages. Her physician said she was not ovulating, a likely sign of low estrogen. Lynn was placed on fertility drugs that brought about ovulation, enabling her to become pregnant and give birth twice, to a boy and a girl. Lynn continued to have irregular menstrual cycles until age fifty-four, when she became menopausal. Her doctor measured her bone density when Lynn was sixty, which confirmed that she had severe osteoporosis, putting her at great risk of bone fractures. However, due to a family history of cancer, she was fearful about starting estrogen.

Lynn's medical history told me that she probably had had low estrogen levels for most of her adult life, which was one probable reason why menopause had such a profound effect—not only on her bones but also on her brain. The key question: Was the effect on her brain one of early Alzheimer's, or a memory impairment that would more readily respond to hormone treatment? Lynn and I had a long discussion about her cancer fears, and I suggested that in her case the benefits of estrogen treatment on her bones and brain might outweigh the cancer risk. I reassured her that we'd carefully monitor her with frequent mammograms. She agreed, and I started Lynn on a moderate dose of estrogen.

When Lynn walked into my office six weeks later, she had a broad grin on her face. "I can't believe it," she said. "My memory is coming back. Does this mean I don't have Alzheimer's?" Fortunately, I was able to inform her that I did not think she had Alzheimer's, and that her memory would gradually improve the longer she remained on estrogen.

It has been two years since Lynn first began estrogen, and while she still has some memory lapses, they are infrequent and mild—the kind that would be expected with normal aging. No longer does Lynn forget something as important as picking up her child, and her fears about Alzheimer's have disappeared.

THYROID HORMONES AND MEMORY: THE NEGLECTED FACTOR

Experts in endocrinology and neurology have known for years that marked hypothyroidism can induce neurological and mental dysfunction.[6] What is less well known is that memory loss can be a symptom not only of severe thyroid dysfunction, but of mild dysfunction, as well. The patient with mild hypothyroidism (low thyroid hormones) is usually free of neurological complaints, and laboratory testing of thyroid hormone levels show only a slight deviation from normal.

One of the most common causes of hypothyroidism is a genetically predisposed disease called Hashimoto's thyroiditis. It is much more prevalent in women, particularly in the age range of menopause, when its incidence is five to eight times that of men of a similar age. Hashimoto's thyroiditis is an autoimmune disease, in which the immune system attacks the thyroid gland, and is characterized by the presence of circulating antibodies in the blood. Since memory loss is not widely recognized as a symptom of hypothyroidism, the diagnosis is frequently missed. And the fact that it is more prevalent in older women explains why it is sometimes mistaken for Alzheimer's disease. It's important that doctors differentiate between Alzheimer's disease and hypothyroidism, especially since the memory impairment of hypothyroidism can be effectively treated with thyroid hormone.

In one study, patients with hypothyroidism were given a test of memory and shown to have a significant impairment, as compared with a matched group of healthy adults.[7] After treatment with thyroid hormone, however, these patients showed a marked improvement in their memory skills. Like estrogen, thyroid improves only specific types of memory functions. In this particular experiment, the hypothyroid patients scored poorly on only one subtest of memory, in which subjects were asked to read and then recall details of a story forty-five minutes later. Before starting on thyroid hormone, patients had great difficulty recalling specific details; after treatment, they were able to once again remember and recount the minor, colorful highlights of the stories they read.

Why is the thyroid so important for memory? As with estrogen and testosterone, thyroid hormones appear to uphold the integrity and numbers of nerve cells in central memory-processing areas of the brain. Portuguese researchers conducted animal studies in which they proved that low thyroid levels resulted in a significant loss of cells—as much as 23 percent—in key regions of the hippocampus.[8]

CYNTHIA'S STORY

A small but significant subset of my patients with memory problems have low thyroid hormones due to Hashimoto's disease, other forms of autoimmune thyroiditis, or hypothyroidism with different causes. Cynthia, fifty-six, was one such patient. As a real estate agent, she was always having to describe properties to her clients. "I was telling a young couple about a small but attractive ranch house," Cynthia recounted. "I wanted to say that the upkeep would be economical, but for the life of me I couldn't find the word 'economical.' I didn't want to say 'cheap.' I had a picture of the word in my mind, but it was out of focus and I couldn't see it. I really became concerned when I realized that this memory problem was affecting my work. All of a sudden I felt old, and somehow diminished."

Cynthia had first come to see me two years earlier. I'd prescribed estrogen for other menopausal symptoms, and she did well for quite a while. I had checked her thyroid at the time, and while her hormone levels were normal, thyroid antibodies were present in her blood. Since patients with thyroid antibodies frequently develop hypothyroidism, I closely monitored Cynthia. When she began being plagued by memory loss, she made an appointment to see me. "Maybe I'm not getting enough estrogen?" she asked, wondering if her current dose was insufficient. "Is there some way we can tell?" I ran another battery of hormone tests, and her blood estrogen was perfectly normal, but her thyroid hormone studies showed that she had, in fact, developed hypothyroidism. I was able to reassure Cynthia that her low thyroid level was the likely cause of her memory loss, and that treatment with thyroid hormone should be the solution. After one month of treatment, her thyroid levels were still too low and her memory problems continued. I increased her thyroid dosage, and when I saw her again two months later, she reported a dramatic betterment of her memory.

"I can finally relax again when I talk to my clients," Cynthia said. "I can stop worrying that I will stammer when I'm just trying to remember the right words."

HORMONES AND MENTAL SHARPNESS

Memory is not the only aspect of cognition that is influenced by hormones. So is decision-making and overall mental clarity, as I mentioned at the start of this chapter. Low or fluctuating hormone levels are a common cause of mental fogginess, verbal slips, and an inability to concentrate.

LOUISE'S STORY

Louise, a psychotherapist, was fifty-nine years old when she came for a consultation. She'd become menopausal at fifty-one, and at first she had no severe symptoms of "the change." Since her mother and maternal aunt had breast cancer, she decided not to go on estrogen replacement. In time, however, Louise's energy flagged and she had trouble sleeping. But her worst symptom—the main reason she came to see me—was a growing inability to focus and concentrate. This was becoming a serious, embarrassing problem in both her personal and professional life. Louise reported a telling example from her work as a therapist.

"Just the other day, a female patient told me that she and her husband were having serious marital difficulty related to his alcohol abuse," said Louise. "The situation was so bad that the woman was considering divorce. About fifteen minutes later, she referred to this again; I had to ask her to repeat some of the story. I had been distracted and unable to focus while my patient was talking. She looked perplexed, asking me point-blank if I had forgotten what she'd told me. It was an embarrassing and frustrating moment." Louise said that she'd never had this problem in the past, but in recent times it was happening more and more. "I knew that I was suffering from cognitive dysfunction, and I began to wonder if it was related to menopause."

Her story is typical of cognitive impairments experienced by some menopausal women. Another symptom that often appears alongside this inability to focus is difficulty with decision-making. This problem can involve the most mundane of everyday choices. Louise had tears in her eyes when she told me that supermarket shopping had become an intolerable challenge. "I stand at the frozen food section for an interminable period, unable to decide on broccoli, spinach, or beans," she said. Louise had utterly lost confidence in her ability to make such a simple decision.

I asked Louse about her medical history and sexual development. Her puberty had been delayed; she hadn't started having periods until she was sixteen. After that, "they were never regular," she said. "I could go thirty-eight days without a period, and the next time it would be twenty-eight. Sometimes my flow was very light and lasted only a day or two. I had always wanted children but I only could become pregnant once, and I miscarried after three months." She was aware that she'd never had a strong, intense sex drive. At age eighteen she was diagnosed as having hypothyroidism, and her physician prescribed thyroid hormone, which she remained on until age thirty. Her story suggested that Louise had a hormonal irregularity that began at puberty and had continued for most of her adult life.

I talked with Louise about the risks and benefits of estrogen therapy, and she decided to commence with treatment. To reassure her regarding the risks of breast cancer, we set up a monitoring program of regular breast self-examination and annual mammograms. After one month of estrogen treatment, Louise noticed a distinct enhancement in her capacity to concentrate. She was once again able to focus in a professional manner on the needs of her clients. She told me, "I no longer have difficulty concentrating or following my patients' discussions, and I certainly don't have to ask them to repeat themselves." Louise's improvement was typical of what I observe among many menopausal women taking estrogen.

Estrogen and Cognition: The Hard Evidence

In our research during the 1970s and early 1980s, my colleagues and I investigated hormonal influences on "automatized" tasks—abilities such as walking, talking, and typing that require constant repetitive learning before they become automatic. Men with higher levels of testosterone were better at these tasks. But we also showed that women performed automatized tasks most efficiently during their mid-cycle peak in estrogen. It was our first clue to the power of estrogen to sustain mental sharpness.

The question, then, is whether research verifies that hormone treatment improves different aspects of mental sharpness, such as concentration, reasoning, learning, and decision-making. I've already described research that confirms the memory-enhancing power of estrogen. While there is more evidence on the memory front, several studies do support hormone treatments for mental sharpness. (I should also add that there is some overlap between memory functions and other cognitive abilities, such as concentration, a point to which I will return in a moment.)

In several early tests of estrogen therapy for women, researchers reported no significant benefits on cognition. These "failures" probably resulted from a variety of factors. Their testing instruments did not evaluate cognitive functions that are known to be sensitive to hormonal influence. As I have argued for years, the central nervous system is inordinately complex, and some but not all cognitive abilities are measurably affected by hormones. In more recent years, leading experts have homed in on cognitive functions that we strongly believe *are* influenced by hormones.

In Barbara Sherwin's first, fascinating trial of hormones for women undergoing surgical menopause, she gave fifty of these patients either estrogen or testosterone alone, or a combination of estrogen and testosterone, or a placebo. The women taking a placebo suffered setbacks in their short- and long-term memory, but they also experienced a decline in logical reasoning, which includes both concept formation and abstract reasoning. By contrast, women taking any one of the hormone treatments experienced *no* decline in memory *or* logical reasoning, which suggests that mental sharpness may be protected by hormone modulation—especially with estrogen but perhaps also with testosterone.

Dr. Sherwin also points out that her studies of estrogen and memory enhancement included several memory subtests that involved both *repetitive learning* and the ability to *concentrate and pay attention.* Women taking estrogen maintained these cognitive abilities, while those on a placebo did not. Thus, most of Sherwin's research showing a link between estrogen treatment and better memory also suggests that cognitive functions *allied* with memory are also protected or enhanced. These cognitive functions involve learning and concentration.

Both doctors and patients should be aware that not all cognitive behaviors are influenced by hormones. When a menopausal woman complains of cognitive symptoms, it is important for the physician to know which ones may respond to estrogen and which ones may not. Difficulty concentrating or focusing *is* often reversed by hormone treatment. Concentration and focusing requires good automatization, and my own research has shown that this facility is weakened by low estrogen. (For instance, women's automatization ability was poorest during the time in their cycle when estrogen was low.) Therefore, it is not surprising that mental sharpness may falter when estrogen levels are continuously low, as they are during menopause.

The comments I hear repeatedly from my peri- and postmenopausal patients are telling. "I get started on one activity and all of a sudden something distracts me and I'm off on another tangent." "I can't focus on one thing long enough to complete it." "I'm all over the place and nothing gets done." During my early studies of automatization, I didn't really link the ability to automatize with the ability to focus. However, we have since learned, through sophisticated psychological testing, that people who are good automatizers have the ability to ignore irrelevant information and focus on the task at hand. Put simply, when women have sufficient estrogen, they're good at automatic tasks *and* they're better able to focus. These two cognitive facilities frequently overlap.

Hormones and Decision-Making

Another common complaint among my patients is that they can't make decisions. The most frequent is, "I get up in the morning and I'm overwhelmed, because I can't decide what to wear." Such women can take hours trying to put together their workaday or social outfits. I also hear: "I am lost and embarrassed at work because I can't make up my mind." "My kids think I'm nuts because I spend so much time in the supermarket trying to decide what to buy." Louise, the psychotherapist who could not decide which vegetable to select in the frozen food section, was a case in point.

How do we know that patients like Louise are befuddled because their brains are deprived of estrogen? My colleagues, Don Broverman and Bill Vogel, and I planned a study to answer just this question.

In designing our experiment, the three of us had a puzzle of our own to solve: What's the best way to measure this phenomenon? For

Don Broverman, a cognitive psychologist, this was the kind of challenge he enjoyed. A few weeks after our first discussion of this project, Don showed up in my office with a funny little homemade wooden box. As I peered into the box I saw that it contained ten perfectly round cylinders of various heights. Don took out one of the cylinders and placed it on my desk. Next to the cylinder, he piled a stack of twelve homemade wooden poker chips. "How many of these stacks will make up nearly the height of this cylinder?" he asked me. It wasn't immediately obvious, but after a few seconds, I guessed three. Then he said, with a smile, "Are you absolutely sure that's right?" "I'm pretty sure," I hedged. "This," he said, pointing at the box, "is how we can determine whether estrogen affects a woman's decision-making ability, and her confidence in her decisions."

Don and I became instantly excited by his device and the study it would yield. But we were the passionate optimists on our team, and we knew that Don's design would have to pass muster with Bill Vogel, whose exceptionally sharp critical faculties we relied upon. We presented the experimental design to Bill, and were delighted when he shared our enthusiasm.

We evaluated thirty-eight normal menopausal women who had not been taking any hormones before entering the study. We divided the subjects into two groups: The first group would take estrogen for two months, then take a placebo for two months. The second group would do the opposite—take a placebo for the first two months and then take estrogen. When women in both groups got estrogen, they also received progestins for ten days each month to mimic the hormonal pattern of a normal menstrual cycle.

Using Don Broverman's cylinders and chips, we evaluated decision-making ability and confidence twice monthly for each of the four months. During the two months of hormonal treatment, the testing was performed while the women were taking estrogen alone and again while they were taking the combination of estrogen and progestin. (We suspected that the progestins might negate some of the positive effects of estrogen.) During each testing session, the women were asked to judge whether one, two, or three stacks of twelve poker chips would approximate the height of a cylinder. Decisions were to be made about ten different cylinders placed in front of them one at a time. We recorded the accuracy of the decisions, and we used a stopwatch to document the time taken to complete each decision. We also asked the women to rate the confidence with which they made their decisions.

Women in both groups performed this task more accurately, and with greater confidence, when they were taking estrogen. It didn't matter whether they were in the group that took estrogen first or a placebo first—when they were on estrogen, they were better decision-makers.[9]

As we suspected, the women's decision-making accuracy was impaired while they were taking a combination of an estrogen and a progestin. Interestingly, the combination of estrogen and progestin did not seem to significantly impair the women's confidence in their decisions. It was difficult to evaluate the effect of estrogen and progestin on the speed with which the decisions were made, because the speed quickly increased in both the placebo and estrogen groups. We believe this was largely due to a practice effect.

Estrogen's positive effect on mood also appears to act positively on a woman's confidence regarding her decisions. Our testing showed that women who had somewhat dysphoric moods gained more confidence in their decisions when treated with continuous estrogen than women with less dysphoric moods. The addition of a progestin for ten days each month did not negatively affect these women's improving confidence level.

Our study, now being prepared for publication, is to my knowledge the first controlled experiment to document estrogen's positive effect on decision-making ability and confidence. It offers hope to women, whether peri- or postmenopausal, that hormone modulation can resolve their symptoms of indecisiveness and mental fuzziness. These symptoms are no small matter to women who depend on mental clarity in order to lead fulfilling lives at work and at home. I have been delighted, and at times even astonished, by the turnaround experienced by many of my female patients who complain of vacillation, hesitation, doubt, and confusion. Within weeks or months after starting hormone modulation, they undergo a veritable transformation, with a return of mental sharpness and decisiveness that allows them to thrive once again.

PRINCIPLES OF HORMONAL HEALING: STORIES OF RECOVERY

Now that you've learned about hormones, memory, and mental sharpness, I'll describe several cases that demonstrate the comingling of cognitive and psychological symptoms that beset women when their hormones are in flux. These stories also demonstrate how proper treatment involves careful testing and combinations of hormones that work together to optimize mental well-being. I call this treatment hormone modulation. One key to successful hormone modulation for cognitive dysfunction is to find the most effective, least deleterious way to provide progesterone. As my own research has proven, progesterone can offset or eliminate the positive effects of estrogen on mental sharpness. Physicians must develop creative and thoughtful ways to solve this problem, and I will illustrate how it can be solved.

MARY'S STORY: MULTIPLE SYMPTOMS, MULTIPLE TREATMENTS

Mary was browsing in a bookstore when a book fell off the shelf onto her foot. She picked it up and was struck by the title, *Menopause and Madness.*[10] Mary was a fifty-two-year-old menopausal woman who was having trouble navigating the passage, suffering bouts of depression that made it difficult for her to function. The title of the book intrigued her, and when she browsed through it she found that the author, Marcia Lawrence, lived near her, just outside of New York City. Marcia and I had become friends when she called me regarding my research as she started to write her book. Several years before she called me Marcia had suffered a severe breakdown from manic depression. As it happened, Marcia's problems started when her estrogen levels dropped precipitously at the time of menopause and she did not fully recover until she was treated with estrogen. Mary called Marcia, who willingly shared her experiences. The two women spoke at length about how menopause had created havoc in their lives. "She was wonderful," Mary said. "Marcia described her own menopausal difficulties, and she listened to my story with great warmth and compassion." As a result of their conversation, Mary made an appointment to see me.

"I work in the health field," Mary said, "but I can't seem to help myself." Mary had been married for a second time three years earlier, and while she deeply loved her husband, she had lost all interest in sex. During this time, her periods became irregular and finally stopped. Like some menopausal women, she did not have hot flashes or night sweats. Depression and mental fuzziness were her major menopausal symptoms. Mary's physician put her on low-dose estrogen, cycled with a synthetic progestin for ten days each month. But every time she took the progestins her moods worsened. Her doctor prescribed a different form of estrogen, and he switched her to a large daily dose of natural progesterone, taken orally. Again, this combination of estrogen and progesterone made her depression worse. "I lost all interest in life," Mary said. "I was terrified I might kill myself."

Not only was Mary depressed, she had extreme difficulty organizing and focusing her thoughts. "I've always been a little distractible, but this was getting ridiculous," she said. "I'm all right as long as I don't have to make decisions. I had just made an appointment with a new dentist, and the secretary gave me directions to his office. They were pretty simple, but I had to make a number of decisions in sequence in order to get there. I simply couldn't do it. I was totally lost and had to stop to phone the office. And that wasn't even the worst disaster of the day. When I left the dentist's office and went out into the parking lot, I couldn't find my car."

By the time I saw Mary, I knew from our own "decision-making" experiment that estrogen could help her overcome her cognitive dysfunction, and that too much progesterone, or the wrong kind, could make it worse. My first move was to take Mary off daily progesterone. I measured her blood estrogen and testosterone levels, which were low. I, therefore, increased her estrogen dose and started her on low-dose testosterone. (I used the form I rely upon most—methyltestosterone.) I decreased her dose of natural progesterone, and switched her from a daily regimen back to ten days per month. These changes were designed to decrease the negative effects of progesterone on her moods and cognitive abilities.

When I talked to Mary again six weeks later, she reported that her depression had completely lifted. "I can't believe I thought of killing myself," she confided. "Now, I could never imagine doing that to my family." Moreover, Mary was able to focus her thoughts again. "I have no trouble following complicated directions anymore, and I haven't lost my car in a parking lot in a month."

Mary's story is a good example of how estrogen can help some menopausal women regain their ability to focus, remember, and make sound everyday decisions. Her story also shows how mood disturbances and cognitive difficulties often occur together, presumably because low estrogen and other hormonal imbalances affect brain structures and systems involved in both psychological states and mental abilities. Some women's health experts argue that hot flashes and night sweats can disrupt sleep, and that loss of sleep is the real cause of mood and cognitive problems. But we know that the psychological and mental disturbances can occur in the absence of hot flashes and other physical symptoms of menopause—as was the case with Mary.

Mary's case also demonstrates how hormone modulation requires careful attention to blood test results, rectifying the loss of both estrogen and testosterone, and the careful administration of types and schedules of progesterone that do not negate the benefits of estrogen.

JOANNE'S STORY: THE NEED FOR INDIVIDUALIZED TREATMENT

One of Joanne's major complaints was that she could not focus. "The other day I was vacuuming the rugs in my house when I noticed a food stain on the couch," she explained. "I went out to the kitchen to get a wet cloth. While I was there I noticed the dirty dishes in the sink. The next thing I knew, I was filling the dishwasher instead of vacuuming the rug." Joanne's experience exemplifies what happens to some peri- and postmenopausal women: They can't stay on one track at a time. "I cannot sleep and I'm having hot flashes," said Joanne, "but not being able to focus is far worse. I feel like something has happened to my mind, and have no control over it."

Joanne was not only having problems focusing, she was having pronounced mood swings. Her periods had stopped two years earlier, and her gynecologist started her on a standard dose of estrogen, along with a rather high dose of synthetic progestins for fourteen days each month. The estrogen helped her sleep, but during her two weeks on progestins her mood swings worsened and she began to harbor suicidal feelings. Joanne was terrified by this turn of events, and that's when she first came to see me.

Joanne's blood estrogen levels were low, and I put her on a higher-than-standard estrogen dose. (See chapter 8 for specific hormonal medication and the range of dosages.) I switched her from two weeks of synthetic progestin each month to ten days of natural micronized progesterone, which usually has a far less damaging effect on moods and mental symptoms. When I saw Joanne two months later, her moods had stabilized and her ability to focus had returned.

"The other day I was writing out checks to pay the bills when I remembered that I needed to wash out the blouse I wanted to wear the next day," Joanne said. "Two months ago, I would have stopped writing the checks and put the blouse in the washing machine, but I didn't do that. First I finished writing the checks. Afterward, I took care of the blouse. I feel that I'm in control of my mind, and I'm myself again."

Like Mary, Joanne's cognitive symptoms occurred alongside emotional ones, namely severe depression. Unlike Mary, however, Joanne's mental and emotional difficulties occurred alongside the common physical symptoms of "the change," including hot flashes. When she came to see me, Joanne's continuing distress was a result of too little estrogen and an overly high dose of synthetic progestins. The solution, in her case, was more estrogen and a more moderate dose of natural, as opposed to synthetic, progesterone given for only ten days. The resolution of Joanne's suffering helps us to recognize the importance of individualized hormone modulation.

EVELYN'S STORY: CONVENTIONAL SOLUTIONS CAN BE HARMFUL

Evelyn had just turned forty-eight when she began to have problems concentrating. She had no significant problem with night sweats or hot flashes, but she had noticed a decrease in her sex drive, pain during intercourse due to vaginal dryness, and an inability to have an orgasm over the previous six months. Evelyn was a businesswoman, and her difficulty concentrating was becoming a real problem at work. She would attend meetings, and midway through them she'd realize that she wasn't following what was being discussed. "Other thoughts keep intruding, and I end up being totally distracted," said Evelyn. "When I try to focus on what's being said, I realize I haven't been listening and I'm totally lost. This isn't like me at all. I'm very devoted to my job. In the past, I have always been completely involved during meetings. Now, I walk out feeling like I don't have a clue."

Evelyn's gynecologist ordered blood tests, and her FSH levels were high, confirming that she was menopausal. He then placed her on an oral contraceptive, which contained a particular combination of synthetic estrogen and progestin. This is common practice for gynecologists treating women with early menopausal symptoms. In part, I believe, this practice grows out of concern that the patient may not be in frank menopause but rather perimenopausal. During this time, physicians worry that women may still ovulate, raising concerns about the occurrence of an undesired pregnancy late in life. Thus, they reason, the oral contraceptive solves two problems: It treats the estrogen deficiency at the same time that it blocks ovulation. But this logic fails to account for the negative impact on the brain of this combination of estrogen and progestin taken daily. As I have emphasized, the particular combination of hormones in "the pill" is frequently associated with a worsening of mood and an adverse effect on some cognitive abilities, such as mental clarity and decision-making. This is precisely what happened to Evelyn.

Several months after Evelyn began taking the oral contraceptive, she noted mood swings and then a full-scale depression. When she complained to her gynecologist about this, he referred her to a psychiatrist, who prescribed Prozac. But the Prozac did not work, and Evelyn was launched on that unfortunate, all-too-common journey of switching from one ineffective antidepressant to another. When Evelyn came to see me, she was still on that merry-go-round. Her mood was flat and she was having even more trouble concentrating at work. I recommended that she stop taking the oral contraceptive and start a dose of a noncontraceptive estrogen, which I believed would improve her concentration and relieve her depression. In order to protect her uterus against uterine cancer Evelyn was started on oral micronized progesterone, ten days out of each month. After several months of this treatment, Evelyn said that her ability to remain focused had returned full force, and her mood was cheerful and joyous once again.

Evelyn's story is yet another example of the fact that conventional solutions to the hormonal flux of perimenopause and menopause are often inadequate or even harmful. Evelyn's oral contraceptive was clearly doing more harm than good, but she didn't need to go off hormone therapy, she needed appropriate hormone therapy. The problem is that doctors and medical specialists are thoughtful and concerned when it comes to the effect of hormone treatments on the reproductive organs, heart, bones, and breasts. But too many of them don't even consider the effects of hormones—both positive and negative—on the brain. This is overwhelmingly due to lack of awareness, not negligence. Given what we now know about hormones and the brain, it is time for change, not only in our attitudes but in the practice of medicine itself.

4

Harnessing the Chemistry of Desire

SEXUAL HEALTH

When Jane, a retired magazine editor, told her gynecologist about her hot flashes, he was quick to put her on standard hormone replacement—daily estrogens and progestins. Two years later, when Jane complained about her waning libido, that same doctor told her, "It's all in your head." His remark was profoundly disturbing to Jane, who was certain that her problem was physiologic. She felt dismissed, as if her own beliefs about her condition did not matter, and even worse, that her sexual difficulty was either a figment of her imagination or the product of a troubled mind. Meanwhile, this "imagined" problem—a stunning loss of sexual desire—was causing serious trouble in her marriage.

When I evaluated Jane in my office, her blood test results confirmed my initial suspicion: Her sexual dysfunction resulted from extremely low levels of testosterone. Testosterone is the so-called male hormone, the biochemical basis for male sex characteristics, drives, and desires. But women also secrete testosterone, though in much smaller amounts. It is testosterone that stimulates libido in both men and women. Indeed, as Boston psychiatrist Dr. Susan Rako has convincingly argued in her excellent book *The Hormone of Desire, The Truth About Testosterone, Sexuality, and Menopause* (Three Rivers Press, 1999), testosterone does for women what it does for men: It is a major controller of sexual desire.[1]

Consider this conclusion from a recent study of testosterone and female sexuality: "Free testosterone levels were strongly and positively associated with sexual desire, sexual thoughts, and anticipation of sexual activity." Given the conventional wisdom that much of our sexual ability and identity resides in the mind—a wisdom with which I concur—such findings strongly suggest that testosterone feeds the sexual mind.

Many women experience a dramatic decline in sexual desire during the menopausal transition, mainly due to the drop in their testosterone levels. (As I've noted, between a woman's twenties and forties, these levels drop by 40 to 50 percent.) In women, roughly half of their testosterone is produced by the ovaries and the adrenal glands—those three-cornered-hat–shaped glands that sit atop the kidneys. The rest is synthesized in different tissues in the body, although these tissues rely upon precursor hormones also made by the ovaries and adrenals. As the ovaries gradually shut down, they make less testosterone. But the ovaries also provide "signals" for the adrenals to produce androgens, so in essence, the ovaries are like master glands of testosterone. (When women have their ovaries removed, their adrenals make much less testosterone.) In short, as women approach menopause, their testosterone levels drop and this can affect their moods, energy, and sexuality.

When it comes to testosterone deficiency, I must return to the principle that every woman's experience of "the change" is different. While all women undergo a decline in testosterone, not all women suffer resultant mood disturbances, low energy, or the most conspicuous symptoms, sexual disinterest and dysfunction. According to Dr. Susan Rako, "During the two or three years preceding menopause and through five years following, a significant number (roughly 50 percent) of women who approach menopause naturally—that is, with their uterus and ovaries intact—notice some symptoms of testosterone deficiency."

Corroboration for Dr. Rako's statement comes from a Yale University study of menopausal women.[2] In this survey, 68 percent of the women reported sexual problems after the menopausal passage. Of these women, 77 percent said they experienced a decline in sexual desire; 58 percent had vaginal dryness; and 35 percent said that their orgasms were not as intense as before. These numbers tell an important story: Vast numbers of peri- and postmenopausal women experience substantial sexual difficulties. While some women overcome these problems without treatment, I can tell you from my years of clinical experience that many women need help. That means, among other things, treatment with testosterone.

Seven years before the publication of Dr. Rako's book, I prescribed testosterone treatment for Jane. Given her family history of heart disease, I closely monitored her so-called "good" cholesterol—HDL—which testosterone can sometimes lower. But this concern took a backseat to Jane's fears of male-pattern baldness and facial hair. "I was really worried about becoming a bearded lady," recalls Jane. I reassured Jane that these problems would not occur with the low doses of testosterone I would prescribe. The side effects she feared never did materialize, and testosterone modulation worked with remarkable swiftness to change her sexual chemistry. After only one day of treatment, Jane felt a tingling in her breasts and her vagina became more lubricated. Her sex drive was back. Jane and her husband were ecstatic, and the couple have pursued an active sex life ever since.

Jane also has developed a new talent, which began when she started taking testosterone. A lifelong amateur pianist, she began composing music; something she'd never been able to do before. "Not only can I compose, but I can actually go through the tedious process of writing it down," she boasts. "One year, I wrote several songs and presented them to my children at Christmas. Before taking testosterone, I never had that urge." While there is no direct evidence that testosterone fosters creativity, the hormone—which like estrogen facilitates neurotransmission—can boost energy, improve memory, and sharpen concentration. These changes, in turn, may allow patients to tap previously latent creative energies. Jane believes this may be what transpired in her case.

While interest is gaining in the use of testosterone to treat the sexual problems associated with menopause, to this day not all primary care doctors, gynecologists, and psychiatrists are fully aware of how safe and effective this treatment can be. Certainly, for some women estrogen should be part of their treatment for sexual dysfunction, primarily for the vaginal dryness that affects so many. The old term "vaginal atrophy" is inaccurate, even harsh. But for most women, testosterone is the primary hormonal medication to treat the loss of sexual interest, vitality, and the inability to achieve satisfying orgasms.

Jane Brody, the health correspondent for *The New York Times,* recently devoted a column to testosterone for women, "A Tad of Tetosterone Adds Zest to Menopause." It was an excellent, fair-minded, sober yet optimistic assessment of how women may be helped to reclaim their sexual desire and capability with low doses of testosterone. Brody points out that old prejudices including the fear that testosterone will cause "virilization"—growth of facial hair, deepening of voice, loss of scalp hair, and sexual aggressiveness—are exaggerated. Such side effects occur only with higher doses of testosterone that today's knowledgeable clinicians would almost never use. Despite Brody's positive assessment, and some growing pockets of interest and clinical application, many in the medical community have been slow to recognize the potential of testosterone for peri- and postmenopausal women. The shame is that women's sexual suffering, which should be considered a serious condition, too often remains misunderstood and poorly treated. Women rarely get *any* medical treatment for this set of problems, let alone the correct treatment.

Healthy sexual functioning is not a "lifestyle choice," and medical treatments to restore it are not "lifestyle drugs"—some sort of recreational endeavor for women at midlife. For most women, sexual vitality and capacity are *essential* to their physical and emotional well-being. For many, the health of their marriages and partnerships depends, at least partly, upon a vital sexuality. At the same time, sexual health does not always require a partner—many women rely upon masturbation to remain sexually alive and gratified. These issues matter to women, and their medical doctors and health care providers should—must—take them seriously.

If you have sexual difficulties that you suspect are hormone-related, persist and find a physician who will treat your problem effectively. Psychological treatments can be useful, but when hormone deficiency is the root cause of your problem, you need more than counseling—you need hormones administered by a specialist who will listen, take you seriously, and treat your condition, utilizing the latest in scientific knowledge. To accomplish this it is best to arm yourself with scientifically grounded information.

TESTOSTERONE AND SEXUAL VITALITY

Most women experience changes in sexual drive and function during the perimenopausal years and beyond. It is the loss of *estrogen* that leads to vaginal dryness and a thinning of the vaginal wall, all of which can diminish sexual arousal and make intercourse painful. But it's the falloff in testosterone that is primarily responsible for the loss of interest in sex and problems achieving a climax. Thus, it is *normal* for "hormone-challenged" women to undergo some sexual difficulties. Consider the findings of Dr. Harold Persky, a pioneer in research on hormones and behavior. Persky studied two groups of healthy married women whose ages differed by thirty years.[3] As compared with the younger women, the older women had significantly lower testosterone, less frequent intercourse, and they rated their sexual relations as less gratifying.

However, testosterone levels are not the only factor in sexual drive. In a study presented at the 1999 meeting of the North American Menopause Society, it was reported that one of the major influences on sex drive in women is the presence or absence of a partner. Women without partners have significantly lower sex drive than women with a sex partner.

Menopause-associated sexual trouble is not a disease, per se. But it is a condition that women and their doctors should be able to confront without guilt or fear. Together, patient and physician can discuss and solve problems of diminished libido and sexual function. There should be no shame and no judgment—every woman has a right to a full sexual life for as long as she wishes to be sexually active.

While the clinical use of testosterone remains far too limited, a slowly increasing number of menopause specialists are prescribing the hormone to treat sexual dysfunction and disinterest among menopausal women. They are recognizing that too many marriages and loving relationships are harmed or even destroyed—all due to an otherwise treatable hormone deficiency. Not only is a woman's sex drive dependent on normal testosterone, so is her ability to have an orgasm.

One reason testosterone has not been more widely used has to do with the long-ago use of high-dose testosterone, which did result in unwanted side effects—excessive hair growth and other signs of "virilization." As long as forty years ago, clinicians recognized that women's sex drive and function were impaired when their ovaries were surgically removed, and that hormone replacement consisting of estrogen alone did not resolve the problem. Biomedical experts developed medicines that combined estrogen and testosterone, and the combination was successful in treating these sexual deficits. But the testosterone portion of these medications consisted of 5 to 10 mg of methyltestosterone. This was an excessive dose, which too often resulted in acne and facial hair growth, as well as thinning of scalp hair. Consequently, testosterone-replacement therapy fell into disrepute.

I also believe that doctors and patients have been far too reluctant to openly discuss sexual issues as medical problems. Some sexual difficulties do not have a strictly medical cause or cure, but can be identified by a careful history taken by your physician. Psychological complexes, early trauma, and current stresses that cause sexual trouble may require counseling or therapy, not medicines. But many sexual problems are medical issues—at least partly—such as male impotence and ejaculatory disorders, and female difficulties with orgasm, low libido, and other sexual dysfunctions. (If you're not sure whether your problem is medical, psychological, or a combination of both, you should raise this question with your primary care doctor or gynecologist, and when appropriate, your therapist or psychiatrist.)

Women may need both medical and psychological treatment for these conditions, but my point here is simple: Find out if hormones are a cause of your sexual difficulties, and if so, find a physician who will provide appropriate medical treatment with hormones.

The aversion of doctors and patients to frank discussions of sexual conditions remains a stumbling block. Still, there is progress to report, particularly with regard to some doctors' receptivity to testosterone therapy. The interest in testosterone was rekindled in the 1980s, when a number of well-controlled studies demonstrated that menopausal women treated with a combination of low-dose testosterone and standard-dose estrogen had an increase in their sex drive, sexual arousal, and sexual fantasies. The women also had intercourse and orgasms more frequently. Investigators found that these sexual benefits were linked directly to higher blood testosterone—but not to estrogen. These research developments have set the stage for greater openness on the part of some physicians to treating sexual dysfunction with testosterone. I'll return to the research shortly.

What do we know about testosterone's role in sexual desire and function? Research in animals reveals that specific areas of the brain control sexual responses and behavior, and these areas are highly sensitive to testosterone. Some of these areas actually shrink in size if testosterone levels are low and testosterone treatment restores them to their previous size.[4] When little or no testosterone is available to stimulate these brain structures, sexual behavior is literally absent. When researchers implant testosterone back into these specific brain areas, normal sexual behavior resumes.

How does this translate into the realm of human female sexual behavior? Just as men produce small amounts of estrogen, women produce small amounts of testosterone. (Women's bodies make three-tenths of 1 mg each day; men make over twenty times as much, 7 mg per day.) Women's testosterone is secreted by the ovaries and adrenal glands, and it circulates in the bloodstream in two forms. Ninety-eight percent of total testosterone is bound to specific proteins in the blood, one of which is called sex hormone binding globulin—SHBG. The other two percent remains unbound and circulates freely in the blood. This portion is called free testosterone, and is the physiologically active form of testosterone.[5] It is primarily free testosterone that is the biochemical stimulus for sexual desire in both the brain and in the sexual organs.

When menopausal women are treated with estrogen, their sex drive can be affected in two different ways. For some, estrogen boosts sex drive, but for others, it dampens desire. This surprising paradox arises, once again, due to the fundamental truth that menopausal women are not all the same. For example, a menopausal woman whose testosterone levels have *not* dramatically declined may need only estrogen to revive her sex drive. How so? Estrogen is needed to stimulate testosterone receptors in the brain, which allow testosterone to have its stimulatory effect on libido. In such women, the problem is not low testosterone but low estrogen, and the resulting atrophy of testosterone receptors. In these cases, estrogen is often the solution.

Other women experience a *decrease* in sex drive while on estrogen therapy. Among these women, estrogen may stimulate SHBG, the protein in the blood that binds testosterone and makes it physiologically inactive.[6] Due to this increase in binding protein, the level of free testosterone drops down, which causes the loss of sex drive. In recent studies in our laboratory, we demonstrated that estrogen treatment increased women's SHBG, the hormone-binding globulin, by 100 percent. These same women on estrogen had a reduction of free testosterone by approximately 18 percent. (We believe that their free testosterone drops because there is so much more SHBG binding up testosterone in the blood.) Thus, for some women estrogen stimulates a chemical chain reaction that *lowers* free testosterone and therefore dampens sex drive. These women don't need more estrogen, they need more testosterone. As a clinician, I base my treatment decision on certain reliable clues about whether a woman falls into the first category (needing just estrogen), the second category (needing mainly testosterone), or a third category in which she needs *both,* an appropriate dose of estrogen to stimulate testosterone receptors and a low dose of testosterone to replenish a serious deficiency state.

WHAT ARE THE SYMPTOMS?

For women at midlife, the sexual symptoms of estrogen deficiency are well known: vaginal dryness and a thinning of the vaginal wall that make sexual relations more difficult and less satisfying. By contrast, the sexual symptoms of testosterone deficiency have been less publicized, but they are more encompassing. These symptoms include the diminution of sexual energy, drive, vitality, and orgasmic function.

Who experiences these sexual symptoms? As I've emphasized throughout this chapter, any woman who is peri- or postmenopausal. These are women with intact ovaries and other sexual organs who are undergoing the natural decline in ovarian function—and hence, in estrogen and testosterone—that comes with aging. (Remember—a subgroup of women enter perimenopause or even actual menopause as early as their thirties.) Likewise, women who have their ovaries surgically removed may also suffer from a severe decline in sexual interest and capacity. Furthermore, I also evaluate sexual symptoms in women with PMS, postpartum depression, and other conditions associated with a low output of gonadal hormones. The use of the oral contraceptive suppresses the release of the pituitary hormone, LH, which in turn stimulates the ovaries to secrete testosterone. As a result, sex drive and sexual function may be impaired. Usually, discontinuation of the oral contraceptive restores normal testosterone levels.

In *The Hormone of Desire,* Dr. Susan Rako presented a clear, helpful list of six signs and symptoms of testosterone deficiency. Use this as a guideline:

1. Overall decreased sexual desire.
2. Diminished vital energy and sense of well-being.
3. Decreased sensitivity to sexual stimulation in the clitoris.
4. Decreased sensitivity to sexual stimulation in the nipples.
5. Overall decreased arousability and capacity for orgasm.
6. Thinning and loss of pubic hair (in some women).

Dr. Rako properly notes that each woman must assess her own sexual vitality in "the full context of her physical, emotional, historical, and relational circumstances." Try to distinguish problems that are biological in origin—namely, testosterone deficiency—from problems that may be psychological or situational, related to past trauma or present-day stress, for instance. Consider whether your relationship with your current sexual partner is affected by stress, communication problems, or other long-term emotional issues that may influence your sexual desire or ability.

If you find it hard to distinguish whether your sexual trouble is hormonal or psychological, it may be both! For instance, you may be having a relationship problem *and* a falloff in hormones. If that is the case, these two areas of difficulty may reinforce each other. If your sex drive plummets and you can't talk to your partner, the alienation will be compounded. You may need both medical and psychological help. Often, however, a sound relationship is harmed by the loss of sexual desire in one or both partners. In such cases, medical treatment alone may solve the problem.

Dr. Rako has transformed her list of symptoms into seven clear questions you can ask yourself to evaluate whether you are deficient in testosterone:

1. What is my familiar level of vital energy, sense of well-being, sexual desire, and pleasure?
2. Am I suffering a significant loss in this familiar level of energy, of well-being, sexual desire, and pleasure?
3. Do I particularly notice a lack of arousability in my nipples and clitoris?
4. Do I notice not only that I have no particular interest in making love, but also (if this has been part of your sexual life) that I do not feel like masturbating?
5. In even the most conducive-to-me circumstances, does it take a long time for me to be aroused?
6. If I do have an orgasm, is it diminished in intensity?
7. Have I noticed (if this has been a part of your sexual life) a lack of sexual dreams or sexual fantasies?

Many of the women I treat for testosterone deficiency experience problems in all of these areas—energy, desire, arousability, orgasmic capacity, orgasmic intensity, and sexual dreams and fantasies. Others have problems in one or two of these categories. If you have more than one of these symptoms, you ought to be particularly suspicious of a falloff in testosterone.

Dr. Susan Rako makes a trenchant point when she discusses the lack of desire to masturbate as a symptom of testosterone deficiency. In *The Hormone of Desire,* she describes her own loss of sexual interest, and the complete lack of understanding or response from physicians she turned to for treatment. At age forty-seven, while she was still having irregular periods, Dr. Rako was perplexed and profoundly upset by her "significant" loss of general vital energy and sexual vitality. She described it as "sexual deadness," and assumed that she needed estrogen replacement. (This was before she scoured the medical literature and discovered that testosterone was the hormone of desire.) Several physicians refused to offer Dr. Rako estrogen because she was still having periods, a traditional bias that persists in many medical quarters.

In desperation, Dr. Rako sought advice and treatment from a respected male endocrinologist. In trying to describe her sexual "deadness" as carefully and completely as possible, she explained in her book, "Since libido is such a complex phenomenon in relationships, I wanted him to know that I had simple ways of knowing how dead, sexually, I was: My relationship with myself had been pretty dependable since I was a girl." The doctor replied impatiently, "I can't possibly justify recommending hormones because you want to *masturbate*." Dr. Rako felt humiliated, embarrassed, and then angry, "for myself and for all of us who struggle to get the help we need in matters we would prefer to keep private."

Here was a psychiatrist—a physician—belittled as she sought treatment for a condition that was damaging her well-being. (Not that doctors should be treated better than other patients, but when a physician is mistreated in this manner, it's clear that serious prejudices are at work.) Women should be able to have satisfying sex lives, both with and without their partners, for the sake of their quality-of-life and health. Don't diminish the medical and psychological importance of your own sexual vitality, and don't allow physicians to, either. If your doctor does not take your symptoms seriously, find one who will.

THE EVIDENCE FOR TESTOSTERONE THERAPY

A leader in research on hormones and the mind, Dr. Barbara Sherwin of McGill University in Montreal, also spearheads research in hormones and sexuality. With her colleague G. M. Alexander, Sherwin evaluated the links between levels of sex steroids and sexual behavior in nineteen women taking oral contraceptives.[7] The women made daily ratings of their sexual behavior and well-being for an entire month, and they took tests that measured how easily distracted they were by sexual stimuli.

(The more easily distracted, the stronger the sex drive.) During this same period, Sherwin and Alexander tested blood levels of estrogen, progesterone, and free testosterone. The results? The women's testosterone levels were "strongly and positively associated" with sexual desire, sexual thoughts, and anticipation of sexual activity. Moreover, the higher their free testosterone, the more readily distracted they were by sexual stimuli.

While this study confirms the role of testosterone in sexual desire, the question remains as to whether testosterone therapy can rekindle desire in women with low testosterone. Dr. Sherwin has provided answers, with two important clinical studies of testosterone therapy. Her research along with a recent clinical study by Dr. Phillip Sarrel of Yale University comprises strong evidence in favor of testosterone treatment for women with sexual symptoms:

- In Dr. Sherwin's randomized clinical study of fifty-three surgically menopausal women, patients received either estrogen alone, testosterone alone, an estrogen/testosterone combination, or a placebo. Women in the two groups that received testosterone demonstrated an increase in sexual desire, sexual arousal, and sexual fantasies. Women getting estrogen alone or a placebo did not experience such improvements.[8]
- Dr. Sherwin and Dr. Morrie Gelfand evaluated women who had their ovaries removed four years earlier. One group had been receiving injections of both estrogen and testosterone; a second group had been getting estrogen only; a third group received no hormones.[9] Women who took both estrogen and testosterone reported significantly higher rates of sexual desire, arousal, and fantasies than the other groups of women. Moreover, these enhancements occurred when blood testosterone levels were high, but not necessarily when estrogen levels were high. In the two weeks after receiving estrogen/testosterone injections—when blood testosterone was up—these women had more sexual intercourse and more orgasms. "These findings," wrote Sherwin and Gelfand, "imply that androgen (testosterone) may be critical for the maintenance of sexual functioning in postmenopausal women."
- Dr. Phillip Sarrel and his colleagues at Yale University studied twenty postmenopausal women who were dissatisfied with their estrogen-progestin therapy.[10] They were randomized into two

groups: The first group received oral estrogen and testosterone; the second group received estrogen alone. Sexual desire, satisfaction, and frequency were improved significantly by the combined estrogen-androgen therapy, but not by estrogen alone or by the earlier estrogen-progestin combination. An interesting side note: The researchers had measured blood estrogen levels before the study, when the patients were on a standard estrogen-progestin regimen. Later, when these same women took the estrogen-testosterone combination, their blood estrogen levels were not quite as high as before the study—but their sexual functioning was better. This led Sarrel to conclude that testosterone was the hormone responsible for the women's improved sex drive and capacity.

These studies affirm what I have found in my work with peri- and postmenopausal women. When testosterone is given in judicious dosages and proper preparations, it is a safe and highly effective therapy for women with sexual dysfunction linked to the loss of their own natural testosterone during "the change."

SEX, MOOD, AND HORMONE MODULATION

ANDREA'S STORY

Andrea was diagnosed with Obsessive Compulsive Disorder (OCD) when she was still a high school student. She couldn't get a minute of sleep at night until she checked and rechecked every door and window in the house to make sure it was locked. When her periods began, she developed bad bouts of PMS, complete with irritability, outbursts of anger, and frequent headaches. When she entered college, Andrea was hit with an episode of depression. At age nineteen, she became sexually active and began taking an oral contraceptive. Soon after starting on "the pill," Andrea's trouble with depression worsened. Her gynecologist, sensing the link between "the pill" and her deepening depression, switched her to a different contraceptive, which she remained on for five years without severe side effects.

After Andrea got married, she medically terminated three pregnancies. Her fourth pregnancy came to term, but even before leaving the hospital, she sank into a tormenting postpartum depression. Andrea's depression was so severe that a combination of antidepressants and lithium failed to ease her suffering. She was literally bedridden for two years, and it wasn't until she was hospitalized and treated with a combination of lithium and Prozac that her depression began to subside. During Andrea's bouts of depression, her sex drive was virtually nil, and she and her husband Mark endured desert-like stretches of time without sex.

Andrea was thirty-six when she began to have hot flashes, and her doctor placed her on a combination of estrogen and progestin. Thus began a roller-coaster ride of mood swings, during which time she went on and off hormones. Three years prior to seeing me, she switched psychiatrists and began taking Prozac along with a mood stabilizer, Neurontin. But when I first met Andrea, she continued to have serious bouts of depression, and her sex drive was nonexistent. Given her past history and current symptoms, I suspected that Andrea had a long-standing disturbance with regard to hormonal fluctuations and their impact on the brain. Her current symptoms suggested that she was menopausal, and I surmised that her severe reaction to the midlife passage was in keeping with her lifelong hormonal problems.

Blood tests revealed Andrea's elevated FSH, which confirmed my suspicion that she was menopausal. Her low levels of total and free testosterone helped to explain her complete lack of interest in sex. I felt that Andrea was a perfect candidate for hormone modulation. I thought that the standard estrogen-progestin combination was actually worsening her depression and plummeting sex drive, due to the negative impact of progesterone and the insufficient dose of estrogen. I discontinued Andrea's progestin and raised her dose of estrogen. She became agitated for one day after beginning this new regimen, but afterward her condition gradually improved. After several weeks, I added low-dose methyltestosterone to help restore her sex drive.

One month later, Andrea and Mark came for an office visit. For the first time in many years, Andrea's mood had stabilized—she simply was not depressed. She was astonished and grateful at the results after such a short span of time. Not only that, the low-dose testosterone spurred a striking transformation in her sexual interest.

"I haven't felt like this since I was in high school," Andrea gushed. "I can't believe that little pill is doing this for me." Andrea's story illustrates that low testosterone, coincident with other sex-hormone imbalances, can have a devastating effect on both mood and sexual function. Just as surely, her case reinforces the positive impact of testosterone therapy on mood and sexual vitality. Here was a woman with serious, treatment-resistant mood disturbances for most of her life, and when menopause hit, her sex life also suffered. As with many of my patients, hormones were the X factor in Andrea's multileveled disorders, and they were the X factor in her healing.

In 1985, Barbara Sherwin published her placebo-controlled study comparing the effects of testosterone alone, estrogen alone, and an estrogen/testosterone combination on the mood states of surgically menopausal women.[11] Women taking any one of these treatments had a significant drop in depression, with the lowest posttreatment depression in women treated with testosterone alone. While it's impossible to make comparisons between genders, our laboratory and others have demonstrated testosterone's impressive antidepressant effect in men. In the clinical research study conducted in our lab, we found that mesterolone, a testosterone-like drug, was just as effective in the treatment of depressed men as the conventional tricyclic antidepressant amitriptyline.

While low-to-moderate doses of testosterone have antidepressant properties, too much testosterone (or related androgens) can have the opposite effect. In a study of severely depressed premenopausal women at the Worcester State Hospital, my colleagues and I found that blood testosterone levels varied widely across the menstrual cycle. On average, testosterone levels were two to three times higher in depressed women compared to nondepressed women. We believed this to be an important finding, so we pursued this line of research to get further answers.

A complex experimental procedure allowed us to measure the body's total production of testosterone in a twenty-four-hour period.[12] Using this measure, the depressed women produced *four* times more testosterone than nondepressed women. We suspected that the source of this excess testosterone was the adrenal glands. When we administered drugs to specifically suppress secretion of testosterone by the adrenals, their blood testosterone plummeted. This confirmed our hunch that the adrenals were making the extra testosterone. What did this discovery mean? We had also noted that the depressed women's elevations in testosterone were sporadic—sometimes their levels were high and sometimes they were normal. By contrast, the nondepressed women had much more stable testosterone levels. We thought that the depressed women's spiking testosterone was probably caused by bursts of adrenal activity. (Such bursts may occur for a variety of reasons, including stress.) The skyrocketing testosterone—even if temporary—could obstruct some of estrogen's actions in the brain. Most likely, excess testosterone blocks the normal mood-stabilizing effects of estrogen, thus leading to depression.

As with most medications, whether synthetic or biological, too much of a good thing is no longer a good thing. A proper individualized dose of testosterone can counter depression; an excessive dose can worsen depression. Brain health, if you will, results from a balanced biochemistry, which includes a proper balance of endocrine factors—the biochemical bath for the neurons that determine our moods, mental states, and sexual vitality.

THE FACTS ABOUT TESTOSTERONE

Types of Testosterone and Testosterone/Estrogen Combinations

As I've emphasized, the side effects of testosterone become problematic when the dosages are too large. The estrogen/testosterone preparations available in the past contained excessively high doses of testosterone. Some of these combinations contained Premarin (conjugated equine estrogens) with methyltestosterone doses of 5 and 10 mg, which were far too high, and these preparations are no longer on the market. Two lower-dose estrogen/testosterone combinations are now available. Estratest contains 1.25 mg of esterified estrogen and 2.5 mg of methyltestosterone. Estratest H. S. (for Half-Strength) contains half the dose of each hormone. These preparations are very useful for physicians who are reluctant to prescribe a medication that has not been FDA-approved. However, it is also possible to have a compounding pharmacist make up a dosage of methyltestosterone that is considerably lower than those levels of methyltestosterone. Compounding pharmacists are licensed and will make up doses of methyltestosterone that are specifically designed to treat the needs of individual patients. I do not particularly like to prescribe fixed-dose estrogen/testosterone combinations, since they do not allow for flexible, individualized hormone modulation. For that reason, I have prescribed methyltestosterone compounded by a qualified pharmacist in doses ranging from 0.3 to 1.0 mg. If some patients require higher doses, I do not hesitate to administer them, but the large majority of women respond well to low-dose methyltestosterone. Table 3 is a list of the testosterone preparations, some of which I utilize.

What about testosterone injections? Testosterone is available in an

TABLE 3 *Testosterone and Testosterone/Estrogen Preparations*

Trade Name	Generic
Estratest—oral	esterified estrogen 1.25 mg/ methyltestosterone 2.5 mg
Estratest H.S.—oral	esterified estrogen 0.625 mg/ methyltestosterone 1.25 mg
Compounded Methyltestosterone—oral	methyltestosterone 0.3–1.2 mg
Depo-Testosterone—injectible	testosterone cypionate 200 mg/ml
Delatestryl—injectible	testosterone enanthate 200 mg/ml
Depo-Testadiol—injectible	testosterone cypionate 50 mg/ml estradiol cypionate 2 mg/ml

injectable form, but I do not prefer this route of administration for women, since it is usually administered every two to three weeks, which can create a problem. Twenty-four hours after the injection, blood testosterone levels reach peak values that are well above the physiologic (normal) range. Within ten days, blood levels often plummet to subnormal levels.[13] The result is a wide fluctuation in blood levels of testosterone that can produce swings in libido, mood, and energy.

Find a physician who is flexible enough to consider low doses of methyltestosterone that are modulated ("titrated") to meet your individual needs based on your symptoms, and who will monitor your progress. The monitoring of menopausal women receiving methyltestosterone (MT) is made somewhat more difficult by the fact that there is no clinical test available to measure blood levels of MT. Blood total testosterone levels do not reflect MT levels. When methyltestosterone is absorbed from the gut, it travels to the liver, where almost half of it is metabolized to be excreted from the body. Dr. Susan Rako in her book *The Hormone of Desire,* describes in detail how MT is handled by the liver. The liver acts to remove the methyl grouping from most of the MT. The testosterone, both with and without the methyl group, circulates in the blood, bound to carrier proteins. A small percentage of these two types of testosterone is free (unbound by proteins) and will attach to testosterone receptors. The testosterone with the methyl group has less affinity for these receptors and is less clinically active than the free testosterone, which contains no methyl grouping. With the current clinically available methodology, it is not possible to measure methyltestosterone. It is possible to measure free testosterone. Free testosterone levels are increased by the administration of MT. While it is not the most accurate measurement of testosterone activity in a woman who is taking MT, I find it useful. As free testosterone levels rise with the administration of MT, I am sure that the MT is being absorbed from the gut.

The goal of monitoring is to make sure that the patient achieves physiologically normal free testosterone levels, and that her symptoms abate. Her mood, energy, and sexual vitality may benefit from just the right dose of testosterone.

New ways of administering testosterone to women are being developed and tested. Results of a multicenter study published in the *New England Journal of Medicine* in September 2000 reported on the effects of transdermal testosterone (skin patch) in a group of seventy-five women ages 31–56 years who had impaired sexual function following surgical removal of their ovaries.[14] Since the ovary is a major contributor to a woman's testosterone levels, these women all had low blood testosterone levels prior to treatment. Transdermal testosterone improved sexual function and psychological well-being.

Side Effects of Testosterone

"Am I going to grow a beard?" This is a very common question when I discuss the use of testosterone with menopausal women. Many women are totally unaware that testosterone is circulating in their bloodstream, and are fearful of using a male hormone. I reassure them that the side effects of testosterone are generally only a problem when the dose administered is far too high. Excessive doses of testosterone can produce unwanted facial hair and acne.

Testosterone can also have emotional side effects. Perhaps the most common complaint is irritability. Recently, one of my patients, Sandy, told me that she was so irritable after starting testosterone that her four sisters stopped speaking to her. I assured her that her irritability would probably decrease soon after she became biochemically "adjusted" to taking testosterone, but that if she remained irascible we could certainly lower the dose. As it happened, in Sandy's case I did not need to alter her dosage.

Other women are concerned that they will become too assertive. I started Grace on testosterone because she was suffering from a low sex drive. The treatment boosted her libido, but she was worried that it was harming her relationship with her husband. "We get into a disagreement and I can't let go of it," she remarked. "Normally these things roll off my back, but now I can become really stubborn." Even though testosterone was having a positive effect on her sex drive, Grace eventually decided to discontinue the hormone. In my experience, this problem is quite rare. I have had a small number of women in whom even a low dose of testosterone resulted in depression. The most likely explanation for this side effect is testosterone's antiestrogenic action in the brain. In such cases, an increased dose of estrogen may offset the negative impact of testosterone. If this doesn't work, testosterone may have to be discontinued. Two other side effects that have been reported are intense dreaming and physical hyperactivity. I have not observed these problems in my patients, most likely because I rely primarily on low doses of testosterone.

In my experience, testosterone should be given initially in low doses, and increased only if the desired effects are not achieved. When women are guided by their clinicians to take the proper doses of the best available preparation of testosterone—mainly oral methyltestosterone—they will not experience untoward side effects. Doctors who are flexible in their dosing strategies, and who follow their patients' progress carefully with office visits and blood tests, will enable women to overcome their symptoms of low energy, slouching sexual vitality, and sexual dysfunction. One other concern—a putative link between testosterone and heart disease—should be considered by women as they embark upon testosterone therapy.

Testosterone and Your Heart

Heart disease is the leading cause of death in menopausal women. Forty-eight percent of menopausal women will eventually die of heart disease.[15] When one looks at heart disease mortality curves according to age in women, you see a sharp rise during the menopausal years. Women appear to be protected against heart attacks, to a large extent, prior to "the change"—when estrogen is high. Estrogen is a proven protectant for the heart: When menopausal women are treated with estrogen, the incidence of heart attacks is reduced by 40 to 50 percent.[16] While it's clear that estrogen is good for the heart, the relationship of testosterone to cardiovascular disease remains unclear.

A heart attack results when a clot is formed in an artery carrying blood to the heart. The oxygen supply becomes restricted, causing the death of a portion of heart muscle tissue. The clot typically occurs at a place in the coronary artery where cholesterol has been deposited over time, creating a "plaque." Some studies indicate that testosterone may help to guard against heart attacks in men. Testosterone administered to men produced a clot-dissolving effect, thus preventing the formation of the clots that lead to "myocardial infarction"—a heart attack. Cholesterol plaques seem to occur at the site of arterial injury, and testosterone appears to protect the blood vessel lining from injury.

What about women? Studies of female primates confirm that estrogen dilates the arteries of the heart, suggesting a preventive effect of estrogen on heart disease. The addition of methyltestosterone does not diminish the beneficial effects of estrogen.[17] Another concern is the effect of testosterone on blood cholesterol. In women whose ovaries had been surgically removed, Dr. Barbara Sherwin found that the effect of a combined estrogen/testosterone treatment on total cholesterol, HDL—the "good" cholesterol, and LDL—the "bad" cholesterol, was no different than treatment with estrogen alone.[18] In another study estrogen plus 2.5 mg of methyltestosterone was shown to lower HDL levels 23 percent, LDL levels 5 percent and total cholesterol levels by 12 percent.[19] It's not yet clear, however, whether this reduction in HDL has a significant impact on the development of the arterial plaques that lead to heart attacks. The 2.5 mg dose of methyltestosterone is relatively high and in my own clinical practice I have found that I can achieve very good therapeutic results using much lower doses, in the range of 0.3–0.8 mg. With these lower doses the suppression of HDL is considerably less, approximately 10 percent. In some cases I see no suppression at all.

Diseases that cause the ovaries to produce excess testosterone have been associated with an increased risk of heart attack. A relationship between high free testosterone levels and partial blockage of the arteries to the heart has been reported in postmenopausal women with heart disease.[20] Physicians who replace testosterone in menopausal women must take these observations into account. Namely, it's best to maintain normal physiologic levels of testosterone. From what we know today, any increase in heart disease risk is likely to be associated with too much testosterone, not normal levels of testosterone. (When I supplement testosterone in menopausal women, I am usually raising their levels from very low to normal.)

When my patients and I decide together that testosterone is a rational treatment, I routinely start with a low dose that is gradually increased until the desired effects are achieved. Moreover, when I increase testosterone, I make certain to carefully monitor blood cholesterol patterns, including total cholesterol and its good (HDL) and bad (LDL) components. The relationship of testosterone to heart disease in men and women—its risks and benefits—has not been adequately studied. The "more research is needed" maxim holds true here. Currently, only a small number of physicians and gynecologists are prescribing testosterone for their menopausal patients. But it's my belief that we will eventually learn how to use testosterone in a way that protects rather than harms the heart. If future studies confirm and characterize the protective effects of testosterone on the heart, doctors may even save hundreds of thousands of lives by including judicious doses of testosterone in the HRT regimens they prescribe for women.

Testosterone and Progesterone: Do They Counteract?

Unfortunately, it is not widely recognized that standard dosages and preparations used in HRT—which combines estrogen and progestin—can have a dampening effect on women's sex drive. As I've explained, estrogen can lower free testosterone levels by increasing testosterone binding. Also, when you take a typical estrogen/progestin combination, the levels of LH (luteinizing hormone) are suppressed. LH is the pituitary hormone that stimulates the ovary to produce testosterone. Knock down LH, and you knock down testosterone. This effect is especially pronounced with oral contraceptives that contain a powerful combination of estrogen and progestin.

Another estrogen/progestin combination, known as Prempro, can also adversely affect sex drive in some women. Progesterone alone can exert an antitestosterone effect by hooking up to testosterone receptors—thus blocking them from receiving your own testosterone. A portion of the chemical structure of progesterone resembles that of testosterone. Progesterone acts like a key that fits into a lock but can't open the door; not only does it get stuck there, it prevents the real key from entering and turning the cylinder. That's why women frequently complain about a lack of sex drive during the ten to fourteen days when they are taking their monthly cyclic progestin. One of my patients, Helen, repeatedly complained to me that she had difficulty achieving an orgasm while she was having sex with her husband. But her problem occurred only during those days when she was taking progestin. "Normally, I'm able to climax whenever I want," said Helen. "But when I'm taking my progestin, I lose that control, and it can take me forever. Sometimes I just can't climax at all. It's incredibly frustrating for my husband and me; we can both get absolutely exhausted."

Such women must find alternative ways to take progesterone. As I will detail in chapter 7, oral micronized progesterone does not always have the negative impact on mood, cognition, and sexuality that synthetic progestins have. If these oral preparations do not solve the problem, vaginal suppositories of natural progesterone can protect the uterus from cancer with little systemic absorption, and produce less of the negative effects on sexuality.

Other Androgens

Androgen is a generic term referring to the family of "male" hormones that includes testosterone. (I must always put male in quotes, since women make androgens just as men make estrogens.) The other key androgens are dihydrotestosterone (DHT), dehydroepiandrosterone (DHEA), dehyroepiandrosterone-sulfate (DHEA-S), and androstenedione.

DHT is an active metabolite of testosterone, and it's the androgen that stimulates the growth of the male external genitalia in fetal development, and at the time of puberty. DHT and testosterone are the only potent androgens capable of reacting with androgen receptors. Neither DHEA, which is primarily secreted by the adrenal glands, nor androstenedione, which is secreted by the ovary, has strong androgenic or masculinizing effects. However, in women DHEA and androstenedione serve as precursor hormones to testosterone.

Two thirds of a woman's blood testosterone derives from the conversion of DHEA and androstenedione to testosterone. DHEA and androstenedione are both available over-the-counter as supplements. When I have examined women who've been using DHEA as a supplement, I found elevations in their blood testosterone levels. Similar effects are known to occur in male athletes using androstenedione for its performance-enhancing effects. For this reason, androstenedione is illegal in many sports. (You may recall the controversy when it was revealed that over-the-counter androstenedione was being taken by the St. Louis Cardinals ballplayer Mark McGwire, who broke the single-season home run record in 1998. It was and remains legal in baseball.) These hormone preparations are unregulated by the FDA, and we don't have sufficient information regarding their long-term safety. Until we do have more knowledge of their risks and benefits, I believe it is unwise to use them.

Testosterone and Vital Energy

As Dr. Susan Rako makes clear, the primary symptoms of women with testosterone deficiency involve the loss of sexual vitality and function. All too frequently, however, women who lose their sexual spark also lose their vital spark—they are simply exhausted all the time. I can no longer count how many women tell me that their energy and sexuality have both gone "out the window." These women seem to drag themselves through their days, without much excitement, but with a sexual "deadness" or "flatness," and a dearth of physical and emotional energy. Some women's primary problem is utter exhaustion.

NANCY'S STORY

Fatigue had become a way of life for Nancy, who struggled each day to get out of bed. There were many days when Nancy convinced herself that she was seriously ill, and stayed home all day, firmly planted under the sheets. Nancy had an impressive mind and equally impressive credentials. She loved her job teaching at a small college in New York City, but now it seemed like an overwhelming burden. She had experienced similar feelings and sensations in the past, during brief bouts of depression. Now, she wasn't depressed, but she had absolutely no vital energy, which turned everything she did into a monumental challenge.

In the midst of her difficulties, Nancy read Marcia Lawrence's book, *Menopause and Madness,* the story of the author's devastating decline into psychosis during her menopausal passage. Hormone modulation enabled Marcia to recover, and she credited me in her book. Nancy wondered if I might be able to help her, so she wrote me a letter outlining her medical history. I found the letter both revealing and poignant. Here is an excerpt:

> During the perimenopause which started six years ago I attempted to deal with my mild hot flashes and low energy with various herbs, as teas and tinctures. My menopause occurred a year and a half ago when my endocrinologist/gynecologist who has spent her career researching estrogen, found my FSH was above forty. [Indicating the onset of menopause.] She started me on the Vivelle patch (estradiol 0.075 mg), and then on vaginal progesterone. I continued to feel really bad: humorless, detached, distant, hopeless, empty, I felt as if it took all my efforts just to hold myself together each day. I have lost all my drive. I don't have ideas surging through my mind when I wake up in the morning like I always did. Certain alternative quarters advised that accepting the changes that have befallen me would lead to growth and wisdom, but I don't buy this. I would like to have the capacities, the excitement and aliveness that I had before. Do you think you can help me? I am wondering half seriously if I might have a brain tumor. My physician specializes in the gynecological aspects of menopause and HRT. The effect of hormones on the brain is not her main interest and this has been problematic for me.

I was very impressed and moved by Nancy's letter. Clearly, she was a sensitive, highly intelligent woman in terrible distress. I strongly suspected that Nancy needed changes in her hormonal treatment. She came to my office, and blood testing confirmed that her estrogen levels were low. I was certain that an increase in estrogen would help Nancy, so I switched from the estrogen patch to a higher dose of oral estrogen. (On retesting, Nancy's blood estrogen remained low, so I raised the dose once again—a common adjustment in the process of hormone modulation.) While some of Nancy's troubles were caused by low estrogen, her blood work also confirmed a testosterone deficiency, which I felt was the key to her lack of energy. I started her on oral methyltestosterone.

After one month, Nancy came for a return visit. She could not believe the boost in energy, which had gradually returned to near normal since her last visit. Two weeks later, Nancy called to report even greater improvement—the excitement in her voice was music to this clinician's ears. Now my only concern was to find a form of progesterone for Nancy that would protect the uterine lining from cancer without sacrificing the benefits of estrogen and testosterone on her vital energy. After we discussed the possibilities, she decided on a vaginal progesterone suppository for ten days each month. This form of progesterone worked well, and it had no serious side effects. Nor did it offset the gains she made.

When I saw Nancy two months later, I noticed a difference in her appearance. I thought she had been exercising. "Now I have the energy to work all day and still walk two miles," she said. "I've lost some body fat and traded it for better muscles." Nancy had experienced the anabolic effects of testosterone. Testosterone stimulates the growth of lean muscle mass. When the muscle mass is increased, it often replaces fat and more calories are burned during both rest and exercise. Seven times more calories are burned by a pound of muscle compared to a pound of fat. Fat is burned in order to fuel these new muscles and significant weight loss can occur. Testosterone fuels both the growth and strength of muscles, which also helps to escalate energy. Nancy thanked me, but in the final analysis, I had not restored her vitality. Testosterone had done the job.

A Whole-Person Approach to Sexual Vitality

If you have the impression that testosterone is the centerpiece of my approach to treating midlife women with sexual symptoms, you're right. But while testosterone is unarguably the hormone of desire, there is no doubt that estrogen plays a role in sexuality, mainly by keeping the vaginal wall from thinning and maintaining proper lubrication. Estrogen creams applied locally to the vaginal area may be extremely helpful for women with vaginal dryness, mainly for women who have vaginal atrophy but are fearful of oral estrogen. Other hormonal imbalances, including low thyroid, can also be part of the pathological picture when women's sexual health is compromised. Abnormally low thyroid function can affect the secretion of hormones by all the hormone-producing glands in the body. Low thyroid levels decrease hormone levels, including estrogen and testosterone. Thus, while testosterone is the hormonal star when it comes to sexuality, other hormones play key roles. Overall hormone modulation is therefore the best endocrinological approach to sexual difficulties among women before and during menopause. Together, clinician and patient must look at the whole picture, not just testosterone.

Even if your sexual problems are caused by hormone imbalances, you should consider other dimensions that contribute to your sexual difficulties. Psychological issues, relationship conflicts, chronic stress, and a poor diet can be factors—either in the presence or absence of concurrent hormonal problems. Evaluate all these realms as you explore the reasons for your sexual symptoms. Be honest with yourself, your partner, and your doctor. It can never hurt to consider the stresses in your life, and to pursue stress management, relaxation techniques, and, if needed, counseling or therapy for problems with your partner. I have no doubt that hormone treatments are the primary medical solution for countless women with sexual dysfunction, but I'm equally certain that women benefit by optimizing their emotional, nutritional, and physical health on every level. Take care of yourself in all of these dimensions, and you can reclaim your sexual health and well-being.

MEN, HORMONES, AND SEXUAL DYSFUNCTION

The andropause (male menopause) frequently goes undiagnosed. The symptoms are usually less acute than those of the female menopause, but over time, men can become just as adversely affected by their declining hormone levels as women. One of the most dramatic changes can be the loss of libido. To some extent, men's reduced sexual desire may be related to their loss of sexual function. Many men have trouble acknowledging their difficulties, but in my office I speak with men who are willing to admit their problem and acknowledge their feelings about it. They often say that their erectile function is compromised and their sex drive has begun to flag. They have difficulty achieving a firm erection sufficient for sexual intercourse, and even if a partial erection does occur, they often can't maintain it long enough to achieve a gratifying mutual experience with their partners.

Obviously, men's sexual difficulties during and after the male menopause can cause serious trouble in their partnerships or marriages. As frustration grows for each partner, the only "solution" may be to stop trying altogether. But this is no real solution, and it usually breeds depression, anger, or both. Some relationships are strained to the breaking point.

Not infrequently, women and men experience flagging desire or sexual dysfunction at the same time. In these instances, women need to understand men's sexual difficulties so they don't put the onus on themselves. For instance, if you're a woman whose libido has declined and you seek successful hormone treatment, your sex drive may come back to life. But if your male partner is also having hormone-related sexual difficulties, you and your partner will have continuing problems as a couple, and your own efforts may still yield more frustration. Studies have shown that women's continuing sexual vitality depends not only on hormone levels, but on having an ongoing sexual relationship. The "use it or lose it" philosophy is not the whole story, but it's clearly an important biological and sexual truth.

What's clearly needed, then, is proper hormonal diagnosis and treatment for both partners, and a great deal of compassion from each partner for the other. The goal should be for both members of the couple to take rational steps to revitalize their sexuality individually, and together.

Testosterone and Men's Sexuality

In healthy aging men, the downward spiral in sex drive and function is often due to decline in testosterone levels. Men experience a 40 to 50 percent drop in blood levels of both testosterone and estrogen by the time they've reached the age of sixty. Both testosterone and estrogen (estradiol) in the blood are bound to a protein called sex hormone binding globulin (SHBG). Only the unbound (free) hormone is physiologically active, exerting effects on muscle, bone, and brain. Blood SHBG levels increase with age, more testosterone and estrogen are bound, and the result is a significant dip in blood levels of free testosterone and estradiol. These declining free hormone levels result in physical, mental, and sexual changes.

Men who experience sexual difficulties at midlife may resolve them with testosterone replacement. The question is, which men are most likely to benefit? The relationship between impotency and blood testosterone is not clear-cut. Not all impotent men have low testosterone levels. One large-scale study of 1,022 men suffering from impotency found that only 10 percent had low testosterone levels, and in another study of 508 men, only 16 percent had low testosterone. Furthermore, men with low testosterone did not necessarily respond to testosterone treatment. Good treatment outcome was associated with several variables, the most reliable being high levels of LH, the pituitary hormone that stimulates the testicle to produce testosterone. When testosterone goes down, the pituitary gland pumps out more LH in an attempt to jump-start the testicle. This feedback mechanism works only if the testicle is still responsive. In impotent men where the testicle is only partially reacting, LH is somewhat elevated and blood testosterone remains low. These are the men whose erectile dysfunction is most likely to be reversed by treatment with testosterone. When men with this hormonal profile were treated with a testosterone transdermal (skin) patch, they had more long-lasting, firm nighttime erections. Thus, low blood testosterone is only one among several factors that doctors and their patients must consider when developing solutions for male sexual dysfunction.

Testosterone Preparations for Men

There are three different ways to give testosterone to men with sexual dysfunction: an oral preparation, an intramuscular injection, and the transdermal patch. Currently, it is very difficult to sufficiently raise testosterone levels in men with oral testosterone. Table 4 presents the current preparations for intramuscular and transdermal treatment. With injections, blood testosterone levels peak within twenty-four hours and then gradually decline back to baseline over the next seven to ten days. The skin patch gives a more even level of testosterone.

Yet no testosterone preparation or route of administration has been proven to be perfect. In a double-blind study of healthy men with normal testosterone levels who were suffering from erectile dysfunction, high doses of intramuscular testosterone resulted in more frequent ejaculation, but did not influence penile rigidity or sexual gratification.[21] Oral testosterone supplementation administered to twenty-three men with low testosterone levels and impotency resulted in the normalization of testosterone levels in all men but produced a measurable improvement in sexual attitudes and sexual performance in only 61 percent.[22]

My own clinical experience in prescribing testosterone has, in general, been limited to men who have low testosterone blood levels with symptoms of sexual dysfunction, low sex drive, or depression, or all three combined. However, I am cautious about prescribing testosterone for older men who may have a latent prostate cancer. Testosterone may stimulate prostate tumors. Certainly, any older man receiving testosterone must be carefully monitored for prostate cancer with the blood test for prostatic specific antigen (PSA), and regular prostate exams. Also, in younger men who may still wish to father children, intramuscular injections of testosterone should be avoided, since they can suppress sperm counts.

TABLE 4 *Types of Testosterone Preparations for Men*

TRANSDERMAL TESTOSTERONE PREPARATIONS FOR MEN

Trade Name	Generic Name	Daily Dosage	General Use	Brain Effects
Androderm	Testosterone	2.5–5.0 mg	Sexual dysfunction	Antidepressant ↑ Libido and improved cognition
Testoderm	Testosterone	4, 5, and 6 mg	Sexual dysfunction	Antidepressant ↑ Libido and improved cognition
Testoderm (Scrotal)	Testosterone	4 or 6 mg	Sexual dysfunction	Antidepressant ↑ Libido and improved cognition

INTRAMUSCULAR TESTOSTERONE PREPARATIONS FOR MEN

Trade Name	Generic Name	Weekly Dosage	General Use	Brain Effects
Depo-Testosterone	Testosterone cypioniate	100–150 mg	Sexual dysfunction	Antidepressant ↑ Libido Improved cognition
Delatestryl	Testosterone enanthate	100–150 mg	Sexual dysfunction	Antidepressant ↑ Libido Improved cognition

What About Viagra?

The introduction of Viagra—sildenafil citrate—has certainly revolutionized the treatment of male impotency. The erection of the penis involves the release of nitric oxide (NO) into the portion of the penis responsible for erectile functioning. Sexual stimulation causes the release of NO, which then sets off a chain of events increasing blood flow into the two tubes of spongy tissue in the penile shaft, resulting in rigid erections. By bolstering NO, Viagra sustains an uninterrupted blood flow, and hence, firm and prolonged erections.

For many men, Viagra is the solution to impotence, or erectile dysfunction (ED). But several studies indicate that testosterone is required for the synthesis of nitric oxide.[23] A number of men I have treated, men struggling with ED, did *not* respond to Viagra and were found to have a documented testosterone deficiency. However, we do need large-scale studies to confirm that men who don't have an adequate response to Viagra are testosterone-deficient. While Viagra is not currently available for women, clinical trials are under way.

But there is another reason for men to consider testosterone modulation. While ED can often be corrected with Viagra, this wonder drug does not necessarily revive a stalled sex drive. From my clinical experience, many men have a return of their libido while receiving appropriate testosterone treatment. Thus, testosterone alone or a combination of testosterone and Viagra may solve the multileveled sexual problems of aging men.

Helping Men Confront Their Problem

Unfortunately, many men are still unwilling to seek help. When men feel vulnerable in a way that threatens their masculinity, they often choose the route of denial or neglect of the problem. When such men come to my office, I try to create an atmosphere in which they feel safe discussing their problems. I make them aware that they're far from alone in their suffering, which almost always produces a sigh of relief. I recommend to women that they support their male partners in being open and pursuing medical therapy for what, today, is a treatable disorder. Viagra and appropriate hormone modulation with testosterone are the cornerstones of successful treatment for men's sexual difficulties and disorders.

5

Calming the Storms

THE NEW PMS PRESCRIPTION

Seventy years ago, Dr. Robert T. Frank first described the premenstrual tension syndrome (PMS) in terms that would be just as applicable today. For seven to ten days preceding menstruation, Frank explained, his patients complained of tension and irritability, which continued until the menstrual flow occurred.[1] Feeling as though they were jumping out of their skins, they engaged in impulsive actions, in desperately misguided attempts to find some relief. They felt conscience-stricken toward their husbands and families, knowing that they could be intermittently unbearable in their attitudes and responses.

Despite Dr. Frank's lucid description, medicine failed to recognize PMS as a medical entity for many years. Only recently—in the past two decades—has PMS been evaluated as a serious medical problem. Psychiatry has labeled it as "late luteal dysphoric disorder," referring to the fact that symptoms occur after ovulation, toward the end of the luteal phase of a woman's cycle. (The last week or so before a woman's period.) The specific symptoms have been well characterized. Here is the list offered in the current manual of psychiatric diagnoses:

- *Feeling sad, hopeless, or self-deprecating*
- *Feeling tense, anxious, or "on edge"*
- *Marked lability of mood interspersed with frequent tearfulness*
- *Persistent anger, irritability, and interpersonal conflicts*
- *Decreased interest in usual activities*

- *Difficulty concentrating*
- *Feeling fatigued, lethargic, or lacking in energy*
- *Marked changes in appetite, which may be associated with binge eating or craving certain foods*
- *Hypersomnia (oversleeping) or insomnia*
- *A subjective feeling of being overwhelmed or out of control*
- *Physical symptoms such as breast tenderness or swelling, headaches, or a sensation of bloating or weight gain*

How prevalent is PMS? Epidemiologic studies suggest that 30 to 80 percent of women suffer from its symptoms.[2] Three to ten percent have symptoms severe enough to seriously disrupt their lives.[3] According to one estimate, absenteeism from work due to PMS costs American industry ten to fifteen billion dollars each year.

While PMS may finally be recognized and taken seriously, the condition remains poorly understood—even today. Its causes are cloudy, and the media and general public have been prey to numerous myths and distortions about its origins and manifestations. For too long, PMS was considered a kind of emotional weakness or something that women imagined. Today we know better, but biomedical experts still aren't sure whether to classify it as a gynecological, endocrinological, psychological, neurological, or psychiatric problem.

In my view, PMS involves all of these elements. Though mysteries remain, my own research and clinical experience, and several recent studies, suggest that PMS results from the negative effect of hormonal fluctuations—including low estrogen—on brain chemicals and receptors in women who are susceptible. These hormonal fluctuations may be considered "normal," but the effects are not, since these women are vulnerable based on their unique biochemistry, genetics, stress, or a combination of all these factors.

Sound complex? It is, but my message here is relatively simple: Hormonal flux is a key factor in PMS, and for many women with severe symptoms, hormone modulation should be part of the solution. Most women can free themselves from the debilitating and distressing symptoms of PMS with either: (1) hormone modulation alone; (2) antidepressant medication alone; or (3) a combination of hormones and antidepressants. Here's one of the most exciting developments in my clinical practice: Women who feel miserable and helpless because they are not finding sufficient relief from either hormones or antidepressants are finally getting help with a combination of both.

AUDREY'S STORY

Consider the case of Audrey, a thirty-five-year old saleswoman. When she first walked into my office, Audrey was in the midst of her premenstrual phase. She said that she was constantly crying; either unable to sleep or sleeping excessively; and had dropped several pounds in just the past week. She was taking Prozac, which has a good track record with PMS, but it was not easing her depression during those torturous premenstrual days. Audrey's medical and psychological history was one of extreme vulnerability. She began seeing a psychologist at age fifteen, largely because she would go months between periods, and when she did menstruate it was preceded by severe episodes of depression, anxiety, irritability, and anger. When the worst PMS-related anguish hit her, Audrey had thoughts of suicide. Over the previous two decades, she had tried many antidepressants, but they offered little relief, and Audrey had been hospitalized a number of times.

By the time I saw Audrey, she was having regular periods. But blood tests revealed that her estrogen levels remained low throughout her menstrual cycle. As with most PMS sufferers, it was during her progesterone peak following ovulation that her PMS symptoms became intolerable. I suspected that Audrey had never responded to antidepressants due to her low estrogen. (See chapter 2 for a complete explanation of the link between estrogen and response to antidepressants.) Her other hormonal studies, including her thyroid hormones, were all normal. I prescribed a relatively high dose of estrogen, but Audrey's response was quite gradual. In time, however, she reported a noticeable decrease in her anxiety and irritability, especially during the week before menses. The SSRI medication that had never worked before was finally effective—in tandem with estrogen.

Audrey's depression had always been evident on her face, but during her office visits I saw the subtle signs of change: the faint trace of a smile, and a softening of her usual pained expression. After about nine months of treatment, I was totally caught off guard when I asked her how she felt and she replied, "Fine." For Audrey, this represented a monumental change. She reported that she was sleeping soundly and awakening in the morning refreshed and energized. Her depression and anxiety had all but disappeared, and she displayed a serenity I had never seen in her before.

Audrey's case suggests that hormone modulation works for PMS. As I will demonstrate in this chapter, hormone modulation has an important place in the treatment of PMS, but it is far from the only approach. The news here is that hormones are the missing piece in PMS treatment for many women, one that has been largely neglected by primary care doctors, gynecologists, and psychiatrists.

Before I explain hormone modulation for PMS, I will provide a primer on what we know about the syndrome's manifestations and causes. Indeed, our rationale for hormone modulation for PMS is based on our current understanding of its causes, and I believe that patients must be fully informed about every aspect of their condition in order to make wise treatment decisions.

PMS: MANY CAUSES, MANY TREATMENTS

One of the earliest theories about PMS was that it was a progesterone deficiency. Dr. Katharina Dalton, a British physician, was a forerunner who recognized that PMS was a real entity. Thirty years ago, she suggested that a low level of progesterone during the week before menstruation was the hormonal abnormality that caused the myriad symptoms of PMS.[4] In their clinical practices Dr. Dalton and other physicians found that women beset by PMS improved significantly when treated with large doses of micronized progesterone, which seemed to confirm the progesterone theory.

However, a number of double-blind, placebo-controlled studies have contradicted this theory. These studies found that progesterone was no more effective at relieving PMS than a sugar pill. In her excellent book, *Screaming to Be Heard,* Dr. Elizabeth Vliet sets forth her position that a deficiency of estrogen—not progesterone—is the prominent hormonal finding in her patients with PMS.[5] In my own practice, I have made observations similar to those of Dr. Vliet—many of my PMS patients have an estrogen deficiency. Why are these patients susceptible to PMS? In my view, low estrogen can cause a deficit of brain serotonin, which makes the PMS sufferer vulnerable to the antiserotonin effects of progesterone. I will fully explore this theory in a moment, but suffice it to say that estrogen deficiency is *not* the key factor in all cases of PMS.

Various studies show that women with PMS often have normal levels of estrogen and progesterone.

While today's biomedical research is breathtakingly sophisticated, we still can't identify the exact constellation that causes PMS. But we're getting close, with the proviso that this constellation of causes will differ somewhat for different individuals.

The Progesterone Theory

The symptoms begin mid-cycle, after ovulation, and they reach a crescendo in the week before menstruation. Moodiness, irritability, tearfulness, anger, agitation, swelling of the breasts, and bloating are most common. These conditions subside, often dramatically, with the onset of menstrual flow. This sequence of events coincides with the rise and fall of progesterone levels, which is why progesterone has been considered the hormonal villain of PMS. Believers in the progesterone theory point to studies in which women with PMS are given the drug Lupron, which blocks ovulation and keeps progesterone down.[6] Many of these women experience the disappearance of their PMS symptoms. But Lupron also causes a severe drop in estrogen, which can result in depression and anxiety. In other studies, drugs that block progesterone receptors brought about early onset of menstrual flow, but they failed to alter PMS symptoms. All of these studies are hard to interpret, because the drugs used to block hormones bring about a set of artificial conditions. It may seem as though progesterone imbalance is the primary cause of PMS, but if it were, all ovulating women with very high progesterone would suffer with the typical symptoms, and this is not the case. Some scientists believe that abnormal ratios of estrogen to progesterone are the cause of PMS, but most studies fail to support this notion.

The Consensus About Women's Vulnerability

What, then, is the current consensus about PMS? The prevailing theory now is that PMS occurs in vulnerable women who have an abnormal response to the normal hormonal shifts that occur during the menstrual cycle. The key word here is "vulnerable." Which women are susceptible to having these "abnormal" responses? My colleagues and I carried out a study, supported by the National Institutes of Health, to try to characterize this vulnerability. We gave estrogen, and then estrogen plus progestin, to a group of menopausal women being treated with daily estrogen for two twenty-eight-day cycles.[7] During the last ten days of each cycle, a progestin, norethindrone, was added to the estrogen treatment. We used this as a "model" to study PMS symptoms since the pattern of hormone administration mimicked the estrogen/progestin changes occurring in premenopausal women during a normal menstrual cycle. By itself, the estrogen significantly improved the mood of nondepressed menopausal women. When we added a progestin, the positive effect on mood was eliminated in some—but not all—of these women. The question we posed: Who were the women who had bad reactions to progesterone? The answers might tell us something about who is vulnerable to PMS.

Unlike the women who had no adverse reactions to progestin, the vulnerable women had more symptoms of depression and anxiety prior to treatment, and they'd been menopausal for a longer duration. Due to this longer period of menopause, their estrogen levels prior to treatment were far lower. In my view, this relative estrogen deficiency may cause an imbalance in brain chemistry—namely, inadequate neurotransmission. As I explained in chapters 1 and 2, women need sufficient estrogen to enhance the functioning of the neurotransmitters, serotonin and norepinephrine. With chronically low or suboptimal estrogen, these neurotransmitters and their receptors won't do their job efficiently. The result is a susceptibility to mood disorders. So, when you give more progesterone to women with this susceptibility, it can be the proverbial straw that breaks the camel's back. We now believe progesterone's effects become damaging only when there is a preexisting neurotransmitter deficiency.

Thus, women with PMS may be vulnerable to mood shifts and other psychophysical symptoms when their progesterone levels rise during the late luteal phase of their menstrual cycles. Chronic stress, past or present traumas, a personal history of mood disorders, or a family history of a major depression or manic depression may be indicators of the emotional and biochemical vulnerability that make them susceptible to the symptoms of PMS.

A variety of hormonal changes can upset the delicate balance of brain chemicals to trigger PMS. The nature of these hormonal changes are quite diverse. In some particularly vulnerable women, PMS begins with the first menstrual period. Other hormonal events also appear to predispose to PMS, including withdrawal from oral contraceptives that have suppressed normal ovarian function, tubal ligation that can impair ovarian blood supply and result in low estrogen levels, and a hysterectomy in which the ovaries are not removed but have their blood supply impaired by the surgery.[8]

Serotonin Deficiency: A Common Factor in PMS

Whether or not there is a "triggering event," chronically low estrogen is certainly one predisposing hormonal factor. Then again, many women with PMS do not have an estrogen deficiency. Dr. Peter Schmidt, a psychiatrist at the National Institute of Mental Health, undertook a study comparing women with and without symptoms of PMS.[9] Dr. Schmidt concluded that the cyclical administration of estrogen and progesterone only produced a PMS-like response in women who had a history of PMS. Women who had no such history developed no PMS-like symptoms.

My work, combined with others, has suggested a possible explanation. A key insight about PMS sufferers is that they probably have a relative deficiency of serotonin in the brain, which predisposes them to PMS when they are exposed to either their own cyclic estrogen and progesterone during a normal menstrual cycle or in menopausal women when they are placed on cyclic estrogen and progesterone therapy. This possibility has gained credence from several studies that found low blood serotonin levels premenstrually, in women with PMS.[10] The bottom line: disturbances in a woman's neurotransmitter systems or her hormone profile can feed upon one another, setting her up for PMS during that delicate time before menses.

While the serotonin connection is a key to PMS, it is not the whole story—at least not for all women. But serotonin deficiency is certainly a predominant factor, which is why antidepressants that enhance serotonin—such as Prozac, Zoloft, and Paxil—have been used to treat PMS. I have had considerable success treating women with PMS with these medications.

TREATING PMS: ESTROGEN, THYROID, AND ANTIDEPRESSANTS

In this section I will present three case studies that illustrate three different approaches to the treatment of PMS. In the case of Sheila, her PMS needed to be treated with estrogen without an antidepressant. Colleen's

PMS responded to an antidepressant alone, while Linda required a combination of estrogen and antidepressants, just as was the case with Audrey, whom we discussed earlier.

The Antidepressant Treatment of PMS

If we are correct that PMS involves a deficiency of brain serotonin, the logical treatments are antidepressants that specifically correct this imbalance. It is not surprising, then, that the antidepressants with the best track record for PMS are the selective serotonin reuptake inhibitors (SSRIs). Among them are Prozac, Zoloft, Paxil, and Celexa, all of which increase serotonin in the synapse, that minuscule gap between nerve cells. (See chapter 2 for a complete discussion of how SSRIs work.) The boost in serotonin in the synapse improves neurotransmission and relieves depression, anxiety, and irritability—common symptoms of PMS.

A number of large-scale, double-blind, placebo-controlled studies have confirmed the efficacy of Prozac for PMS. These studies, with a total number of subjects exceeding five hundred, showed that Prozac was significantly better than a placebo for relieving the PMS symptoms of tension, irritability, and negative emotional states.[11]

The wide use of SSRIs for PMS have led some to ask: Do patients have to stay on continuous antidepressants to treat a condition that lasts only five or six days out of the month? Therefore, some clinicians have prescribed the SSRI drug Zoloft for just the two-week period following ovulation. Normally, antidepressants take at least three weeks or more to exert their beneficial effects on mood, but in women with PMS the use of SSRIs during the premenstrual period alone appears to relieve the symptoms effectively. I have successfully treated a number of suffering women with this approach.

In my practice, I evaluate each woman to determine which approach is best to treat her PMS. I have found that one particular group of PMS sufferers responds well to antidepressants alone: those with normal levels of estrogen whose distress levels are moderate to severe.

COLLEEN'S STORY

One such patient, Colleen, experienced extreme anxiety during her premenstrual days. When she was twenty-two years old, she had a premenstrual panic attack, and it would not be her last. Colleen awoke in the middle of the night, her heart racing, consumed by a sense of terror. She lay in bed unable to fall back to sleep, listening to the thumping of her heart.

There was no apparent reason for Colleen's overwhelming anxiety. Her life in a small town in Massachusetts was quite orderly. She was very much in love with her husband, Jim, a pharmacist in their hometown. They shared a common interest in nature and were happiest when they were outside hiking or bird watching. Shortly after her marriage, and several months after her first panic attack, Colleen endured an episode of depression. She began psychotherapy, but her depression, and intermittent panic attacks, continued unabated. The panic could become so severe that on some days she would leave her work as a part-time nurse. Since the panic would often occur when she was driving or caught in crowds, Colleen began to avoid those situations. She also lacked energy, to the point where she told Jim to go alone on the nature hikes they'd always enjoyed. Colleen had never before abused alcohol, but she found herself drinking to relieve the anxiety or to help her sleep. PMS was beginning to control her life.

After six months of havoc, Colleen and Jim began to notice a pattern—depression, low energy, and panic attacks that occurred almost exclusively premenstrually. They began to recognize that she was suffering from severe symptoms of PMS. Jim insisted that she seek help, and that is when the couple came for their first visit to my office. Colleen was premenstrual at the time. "This is the worst time of the cycle," she told me. "I'm constantly anxious, and I can only get to sleep if I have a drink." She was taking Xanax, an antianxiety drug prescribed by her primary care doctor. "I don't like taking it," she said. "But sometimes I have to use it twice a day just to keep functioning."

Colleen's blood estrogen and progesterone levels were normal for a menstruating, premenopausal woman. But her platelet MAO levels were definitely elevated. You'll recall that MAO is an enzyme that metabolizes and inactivates serotonin. The high MAO level was consistent with her history of depression and her current diagnosis of PMS—it implies an abnormality in the serotonin system. This abnormality made Colleen vulnerable to PMS symptoms during the normal premenstrual rise in progesterone. Since Colleen's hormone profile was otherwise normal, I did not feel she needed hormone modulation as a primary treatment for PMS. Rather, her therapy had to be geared to correcting her serotonin deficiency. That meant an SSRI antidepressant.

I prescribed Paxil, an SSRI that has both antidepressant and antianxiety activity. Within several weeks, Colleen's premenstrual panic attacks subsided, and her general feelings of anxiety and depression had been alleviated. She no longer needed Xanax to help her ward off anxiety. Her sleeping improved, and her premenstrual headaches disappeared. Subsequently, Colleen complained that her energy was still nonexistent. I found that her blood testosterone was low, and together we decided that a trial of low-dose testosterone would be appropriate. (The testosterone modulation was not her primary therapy for PMS, but a later addition for a more generalized symptom.) When Colleen came to my office two months later, she was glowing. She experienced a boost in energy that enabled her to resume hiking with Jim. They'd purchased an exercise bike and she had begun cycling every day. She'd also secured a job promotion, an additional source of joy and pride.

"My outlook on life has become more optimistic and positive," Colleen told me. "I never before have had a job that was this demanding. I always chose jobs that were below my intellectual capacity, and so I've avoided responsibility. But now, I have the confidence to take this on."

The Estrogen Treatment of PMS

Simply put, I have found that estrogen therapy is useful for PMS sufferers who have an estrogen deficiency. Dr. John Studd, a prominent London gynecologist, has been a major proponent of this approach. In a ten-month placebo-controlled study, Dr. Studd and his colleagues followed fifty patients receiving an estradiol implant—a capsule inserted under the skin that slowly releases estrogen as a long-term treatment for premenstrual mood disturbances.[12] Initially, 94 percent of the patients getting placebo had a positive response, which lasted three months. After two months, however, the estradiol-treated patients had a significantly better response than the placebo-treated patients. Their superior response was maintained for the remaining eight months of the study.

Interestingly, when Dr. Studd and his colleagues introduced cyclic progestin as part of the estrogen regimen, 58 percent of the women began to suffer PMS-like symptoms. (More proof of the role of progesterone in PMS.) These symptoms were partially relieved when the doctors changed the type, duration, and dosage of the progestin treatment. In another study using a transdermal estrogen patch, women's ovulation was suppressed—and their PMS symptoms went away. Here again, the patch did not raise estrogen to superhigh levels—only to levels normally observed during a woman's menstrual cycle. These well-designed studies show that estrogen alone can be an effective treatment for the worst symptoms of PMS.

Estrogen treatment is most effective in women suffering from PMS who have demonstrably low estrogen levels. I have had many patients who were ravaged for years by PMS, and helped little by other approaches, go into a "PMS remission" soon after going on estrogen. When these patients do not experience sufficient relief with hormones alone, the addition of an SSRI antidepressant is often their ticket to PMS relief. More on this approach in the following section.

SHEILA'S STORY

In some cases, estrogen must be combined with other hormones to create a fully successful program of hormone modulation for PMS. Sheila's story is one such example: Estrogen was key, but she also needed thyroid treatment.

At age eighteen, Sheila had her first premenstrual panic attack. It occurred during her first menstrual cycle, after a portion of her thyroid had been surgically removed. Crowds could precipitate Sheila's panic, and soon she became totally agoraphobic, afraid to leave her house and enter public places. Her periods had always been irregular, and at times she could go four months without having a menstrual flow. Sheila eventually overcame her agoraphobia, and she married a young physician.

Following her thyroid surgery, Sheila's doctor had prescribed thyroxine as a thyroid-hormone replacement. In spite of her menstrual irregularities, she had no problem becoming pregnant. But after a complicated delivery, which thankfully did not compromise the health of mother or son, Sheila's panic attacks intensified. During the days before her period, bouts of severe anxiety were accompanied by irritability, anger, fluid retention, and breast tenderness. For a period of time, the antianxiety medication Klonopin helped control her premenstrual panic. By the time she was thirty-seven, however, Sheila's PMS symptoms were wreaking havoc with her life. She consulted Dr. Scott Cutler, an excellent psychiatrist with whom I'd worked on a number of similar cases. Dr. Cutler referred her to me for hormonal evaluation and possible treatment.

I discovered two clear hormonal abnormalities in Sheila's blood workup. Her thyroid-stimulating hormone (TSH) was slightly elevated, indicating she was not getting enough thyroxine. She also had thyroid antibodies, which told me that she had autoimmune thyroiditis—her immune system was attacking her remaining thyroid tissue. Sheila also had a marked estrogen deficiency. These two hormonal abnormalities—low thyroid and low estrogen—were related, and they were the biological basis of her terrible PMS symptoms. Without adequate thyroid, the ovaries are adversely affected, resulting in estrogen deficits. Insufficient estrogen, in turn, can induce a lull in brain serotonin. This lull makes women sensitive to the negative effects of progesterone, and hence, vulnerable to the storms of PMS. Sheila's case points out the domino-like complexities of hormone-neurotransmitter interactions in the brain.

I treated Sheila's two-pronged hormonal problems in two phases. First, I increased her dose of thyroxine. I hoped that correcting her thyroid problem would get her ovaries back on track, producing normal levels of estrogen. This could improve Sheila's brain serotonin, which would protect her against the ravages of progesterone-induced PMS symptoms—of agitation, anger, and panic. However, I knew realistically that this process would take time. Therefore, I placed Sheila on a dose of estrogen, to more rapidly restore the hormone-neurotransmitter balance to her brain. I also had to raise her dose of thyroxine several times before tests showed that her thyroid function had normalized.

Sheila's response to estrogen was dramatic. Previously, her panic attacks had been disabling, and her irritability had threatened her relationship with her husband and three-year-old son. Sheila's improvement was simply remarkable. I watched admiringly as she sat in my office with her young son, able to answer to my medical questions as she gracefully admired the crayon drawing he had rendered of their family. Prior to treatment, this distraction would have elicited an angry rebuke. Once on estrogen, Sheila's panic attacks decreased dramatically—the storms of PMS had subsided. Over time, her normalizing thyroid solidified Sheila's progress, until it seemed her personality had undergone a positive transformation.

A Combination of Estrogen and an Antidepressant

To date, there have been no well-controlled studies using a combination of estrogen and an antidepressant as a treatment for PMS. However, in my clinical practice I have successfully treated numerous PMS sufferers with this combination. When is combination hormone/antidepressant therapy best? It depends a great deal on your blood estrogen level. In my experience, PMS patients with low blood estrogen often respond best when they receive both estrogen and antidepressant. In some instances, I'll treat such PMS patients with estrogen alone, in the belief that correcting their hormonal abnormality may rebalance their brain biochemistry. But if estrogen alone doesn't resolve their symptoms, adding antidepressants often will.

PMS patients with low estrogen who *don't* respond to antidepressants are good candidates for combination therapy. How so? Estrogen stimulates brain receptors for serotonin and norepinephrine; these receptors may have atrophied when the woman's estrogen was depleted. The atrophy of neurotransmitter receptors explains why these women don't respond to SSRI drugs that depend upon healthy receptors. Once estrogen is added, many of these women finally respond to antidepressants, getting relief from the whole range of PMS symptoms.

LINDA'S STORY

Linda's story starkly illustrates the value of combination therapy for PMS. She was thirty-four when she sought my help, two years after she underwent a tubal ligation, a surgical birth control procedure that ties off the Fallopian tubes and prevents passage of the egg from the ovary to the uterus. Linda had always had mild but tolerable symptoms of PMS, but after this procedure there was a devastating change. Her moods swung wildly between depression, anxiety, anger, and irritability. Her energy reserves dwindled, and her inability to tolerate stress forced her to quit her job as a nurse. Even the simplest tasks required enormous effort, prompting her to gulp down six cups of coffee a day just to keep going. Her dark moods and suicidal thoughts led her to seek psychotherapy. In our consultations, Linda's comments about her history led me to suspect that she'd long been suffering from undiagnosed depression. She knew from childhood that her father was severely depressed, though he never sought help. Since her teen years, there were times when Linda used marijuana as a form of self-medication to help her sleep.

Linda was clearly one of those women whose history suggests a susceptibility to PMS. But why had the tubal ligation brought it on full force? In recent years, gynecologists and surgeons have recognized that this procedure can compromise blood supply to the ovaries. Surgeons have since altered their techniques to try to minimize this problem.[13] In Linda's case, the result was very low estrogen levels throughout her menstrual cycle—even though she was still menstruating. I was quite sure that Linda's estrogen abnormality was a result of her tubal ligation, which had impaired her ovarian blood supply.

Linda's gynecologist had tried to relieve her PMS by blocking ovulation, a fairly common strategy. He placed her on an oral contraceptive containing both estrogen and progesterone. However, due to their progesterone content, birth control pills can actually worsen PMS. Such was the case with Linda, whose PMS symptoms were exacerbated, and she discontinued the medication. After six months, her gynecologist referred her to me. I not only detected her chronically low estrogen, but her progesterone test revealed that she was ovulating. There was no question that her symptoms became dramatically worse when her progesterone spiked after ovulation.

The causes of PMS in Linda's case were becoming clear: Her low estrogen compromised the levels and functions of those mood-modulating neurotransmitters, serotonin and norepinephrine. When her progesterone shot up in the days before her period, Linda's neurotransmitter abnormality left her prone to severe PMS. Linda would literally sink into despair, irritability, and serial episodes of anger.

Linda's intense suffering required prompt and aggressive treatment. I immediately prescribed oral estrogen to help restore normal levels. While estrogen would help to correct her serotonin and norepinephrine deficiencies, I felt Linda also needed psychopharmacologic intervention to hasten this process. I suggested Zoloft, which acts specifically to augment the function of brain serotonin. On a combination of Zoloft and estrogen, Linda's moods began to lift, but she would still dissolve into tears or flare into anger prior to her periods. I gradually increased her dose of Zoloft, and in time Linda's premenstrual tears and anger subsided.

Linda's estrogen treatment had several positive effects on the functioning of her brain serotonin and norepinephrine. Estrogen can both increase serotonin production and activate serotonin brain receptors. It lowers MAO enzyme activity, thus slowing the metabolic breakdown of serotonin and norepinephrine. This hormonal brain effect was supplemented by the Zoloft, which boosts brain serotonin. Most physicians or psychiatrists would have simply prescribed an SSRI, the standard approach to severe PMS nowadays. But Linda—and patients like her—would probably not experience a full remission with this one-armed approach. I've seen too many patients with hormonal abnormalities like Linda's who failed to respond to SSRIs alone.

Once Linda had normal estrogen levels and a proper dose of Zoloft, she experienced a complete recovery from severe PMS. I am certain that the combined effect of the hormone and the antidepressant rectified her neurotransmitter dysfunction, permitting her brain to resist the negative actions of progesterone. No longer plagued by progestin sensitivity, Linda was able to go through her menstrual cycle symptom-free.

The combination of hormone modulation and antidepressants is reserved for women with proven hormonal abnormalities and severe PMS symptoms. But there are more women who fall into this category than most clinicians suspect. While I emphasize hormone modulation for mood, cognition, and sexuality in menopausal women, PMS is generally a problem for premenopausal women with hormone imbalances caused by other factors—such as Linda's tubal ligation. However, PMS can be a major problem for perimenopausal women who are having regular periods but whose estrogen levels are dropping. I treat many perimenopausal women in their late thirties and forties whose PMS is associated with declining estrogen. I also treat postmenopausal women on HRT who are having periods—and PMS. For them, fine-tuned hormone modulation and SSRI medication may be needed. For these groups, combination therapy can be a highly effective approach. For some women, it is the only method that works.

THE PMS PRESCRIPTION: MULTIPLE THERAPIES

While it's beyond the scope of this book to present a complete overview of all of the medical and complementary treatments for PMS, I will present a brief review here.

Blocking Ovulation. Some recent studies suggest that blocking ovulation is therapeutic, since it prevents the postovulatory rise in progesterone that's considered the major offender in the turmoil of PMS. For instance, the drug Lupron blocks ovulation and thus lowers blood estrogen and progesterone, reducing PMS symptoms.[14] However, long-term use of Lupron and similar treatments causes a significant estrogen deficiency, which puts women at risk for osteoporosis and heart disease. These may be prevented by the administration of low-dose cyclic estrogen and progesterone in amounts too small to produce the symptoms of PMS.

Oral Contraceptives. One can also block ovulation with oral contraceptives that prevent the rise of progesterone, thereby relieving PMS. But caution is needed here. Oral contraceptives can also worsen PMS symptoms if the contraceptive contains a low amount of estrogen as compared to progestin. A higher estrogen to progestin ratio is optimal, lest the patient suffer from a progestin-related worsening of her symptoms.

Mind-Body Medicine. I encourage all of my patients, including those with PMS, to actively pursue stress management, relaxation, and a sound diet as part of their recovery program. One published study by a group of Harvard researchers demonstrated a 57 percent reduction in severe PMS symptoms among women who regularly practiced relaxation techniques.[15] Psychotherapy may also help those with emotional issues or conflicts that interact with or exacerbate PMS-related mood swings.[16]

Vitamins and Minerals. Nutritional factors have also been implicated in PMS. Studies by Dr. Guy Abraham have shown that women with PMS have significantly low blood magnesium levels.[17] Dr. Abraham theorized that a deficiency in magnesium might result in a depletion of brain dopamine, which in itself may contribute to PMS. Several of the B vitamins have been found to alleviate PMS.[18] In uncontrolled studies, B_6 (pyridoxine) has been found to lessen premenstrual symptoms. The beneficial effects of B_6 are probably related to its role as a cofactor in the synthesis of serotonin and dopamine.[19] B_6 also plays a role in the metabolic clearance of estrogen by the liver. Doses of B_6 in excess of 50 mg daily may cause peripheral neuropathy with numbness and tingling of the hands and feet.[20] I generally recommend that my patients with PMS take a multivitamin or B-complex pill that contains all of the B vitamins.

Nutritional advice is often given to women with PMS, though it is typically based more on clinical experience than controlled clinical trials. This advice consists of certain recommendations regarding lifestyle, such as the elimination of caffeine, alcohol, and tobacco. The advice regarding carbohydrate intake is conflicting. Some doctors recommend a decrease in simple carbohydrates and chocolate and an increase in complex carbohydrates. Other physicians feel that simple carbohydrates may improve the mood. We don't yet have definitive answers, though women can try these different approaches to see which ones help.

Other health care providers recommend a lifestyle change program that features physical exercise. Studies have shown that regular exercise can increase the brain levels of pleasure-inducing and mood-modifying neurohormones, such as endorphins. Exercise also stimulates brain norepinephrine, which improves mood, suppresses appetite, and increases energy. Since exercise is good for women's cardiovascular and overall health, it is a smart idea to develop an exercise program, one that might help with PMS, as well.

Seasonal Changes in PMS and Light Therapy. Seasonal variations may also influence premenstrual symptoms. PMS, which psychiatry has labeled a "late luteal phase dysphoric disorder," because as we discussed earlier in this chapter, it typically occurs during the late luteal phase of the menstrual cycle (after ovulation, and ending with menstruation). This classification makes PMS an "affective disorder," susceptible to seasonal variations. A Canadian study evaluating the seasonal pattern of women with PMS found that 38 percent had worsening of their symptoms during the winter months.[21] One quarter of these women considered their seasonal changes to be marked or severe. Seasonal affective disorder (SAD) responds positively to treatment with bright light (light therapy). A 1993 study from the University of California reported that the depressive symptoms of women with late luteal dysphoric disorder responded positively to light therapy. While more research is needed, it is highly possible that light therapy may be beneficial for women with seasonal worsening of their PMS.[22]

CALMING THE STORMS

I have come to recognize that PMS is a complex disorder with multiple causes, and thus it is not always possible to "cure" all women of all their symptoms. Biomedicine has yet to unlock all of the mysteries of PMS, and intensified research efforts are clearly necessary. But I am convinced that a major underlying cause in PMS is an out-of-balance brain chemistry that is frequently influenced by an out-of-balance hormone profile. These intertwined abnormalities can be successfully treated with the judicious use of antidepressants and estrogen, either alone or in combination. Sufferers should also undertake general measures to enhance their psychological and physical well-being. Every woman's therapy should be carefully modulated to meet her specific needs, to balance her unique biochemistry. By following this prescription, women in severe distress may be able to calm the storms of PMS.

6

The Gift of Consciousness

ESTROGEN AND ALZHEIMER'S DISEASE

THE DEVASTATION OF ALZHEIMER'S

NETTIE'S STORY

Nettie had an identical twin sister, and when the sisters married, the two families bought homes next door to each other. Like most women of that time (early 1900s), Nettie stayed at home and raised her family. She was an excellent seamstress and an accomplished cook. A kind, gentle person, she took great pleasure in doing things with her young grandson Steven, who lived nearby. They shared a special relationship picking blueberries and raspberries in the backyard and making jam together. One summer evening, while Nettie and the ten-year-old Steven swam in the backyard pool, a swarm of butterflies fluttered overhead and she showed Steven how to hold a butterfly in his hand. When Nettie became menopausal, like most women of her generation, she was not given estrogen replacement. Steven was twenty-two when he first noticed that his grandmother, who was now over seventy, seemed quieter than normal. She had always kept her house spotless but Steven began to notice that she was vacuuming and dusting less. Her twin sister, Arlene, also knew that something was happening. Soon thereafter, Steven's grandfather told him that his grandmother was putting food in the oven and forgetting to turn it on. When she did the laundry she would forget to use soap and at other times, the washing machine would overflow with suds from the excessive use of soap. Finally, her conversations with her family became limited to simple, short sentences. Nettie's family became so concerned they talked with the family doctor who suggested they consult a psychiatrist. The psychiatrist administered a Mini-Mental Exam, used to diagnose AD. When Nettie was unable to answer the questions concerning the time of day, the month of the year, and who was president, she became uncomfortable and said to her husband, "I don't like this man; please take me home." Not long thereafter, she was unable to climb the stairs and a bed was put in the den on the first floor. At night, Nettie's disorientation became worse. While trying to fall asleep, she would become upset and tell her husband, William, that someone was hiding under the bed. To ease her fears, William, who himself was eighty-four, slept next to her in a reclining chair. Nettie also became incontinent, and William had to change her clothes as many as six times a day. As happens with most caregivers of AD patients, William became exhausted and had to arrange for home care, but finally the family realized that Nettie required care twenty-four hours a day. A nursing home was found nearby. Sadly, Nettie had lost interest in living and the nursing home was not staffed adequately to fully care for a patient with AD. Under these circumstances, her condition rapidly deteriorated.

With so little positive in her life at the nursing home, the family was pleased when Nettie met another woman there with AD and they became close companions. William, Steven, and other family members visited frequently but as she became less and less active, her health deteriorated, and she became more prone to develop infections. After a devastating year of repeated infections and diminishing mental functioning, Nettie was nearly unresponsive and the family was exhausted. Finally and mercifully, one of the infections led to her death.

ESTROGEN TREATMENT TO PREVENT ALZHEIMER'S DISEASE

Imagine, like Nettie, reaching the twilight of your life and being unable to reminisce with family and friends. Imagine what it would be like to have to rely upon others to meet your basic needs. Imagine that your beloved spouse or parent is stricken with a disease that makes it impossible for them to communicate or take care of themselves, and you have had to take over as full-time caregiver. This nightmare is reality for the estimated 4 million Americans—mostly women—who suffer from Alzheimer's disease.

Many of us have close relatives or acquaintances with Alzheimer's, know someone with the disease, or at least are aware of it because of the increasing media coverage of AD. It has stricken some prominent Americans, most notably former president Ronald Reagan, and we as a society are finally beginning to grapple with the dimensions of this devastating neurological condition.

Alzheimer's is characterized by a form of dementia that includes loss of both short- and long-term memory. The person's ability to remember, think, reason, and coordinate movement is gradually eroded. Neuroscientists have begun to identify the disturbances in the brain that characterize AD, and epidemiologists have joined their neuroscience colleagues to ferret out several key risk factors.

One line of research has yielded a surprising conclusion: Among women, who have a nearly threefold incidence of AD as compared to men, estrogen can protect the brain against the disease.[1] While this conclusion has considerable data to support it, the positive proof that estrogen protects the brain against Alzheimer's disease does not exist. However, there are some data indicating that estrogen treatment can slow or reverse the worst symptoms in at least some patients who already have Alzheimer's.[2] However, there are also data contradicting this. Women with mild to moderate Alzheimer's dementia were given Premarin (1.25mg) or placebo in a double-blind study for sixteen weeks. The results of this study showed that estrogen does not improve the symptoms of most women with AD.[3] In another longer term study women with mild to moderate AD were treated with two doses of Premarin (0.625 or 1.25mg per day) or a placebo. At the end of one year there was no significant difference between estrogen and placebo groups in any of the measures used to evaluate the symptoms of AD.[4]

While estrogen therapy has been reported to increase brain blood flow in menopausal women, a study in the *Journal of Neurology* in June 2000 reported no effect of a twelve-week trial of Premarin (1.25mg) on cognitive performance, severity of dementia, mood, or brain blood flow in female Alzheimer's patients.[5]

While these studies seem to be discouraging in regards to the use of estrogen for AD treatment, Carlson and Sherwin found that mood symptoms in a group of women with Alzheimer's were related to the blood estrogen levels.[6] The group was divided into estrogen users and nonusers and as expected higher estrogen levels were found in the treated women. Subjects with higher blood estrogen levels had fewer mood symptoms compared to women with lower blood estrogen levels. The highest depression scores occurred in Alzheimer's patients who were estrogen nonusers. The depression scores of women with AD were significantly greater when compared to scores of women without dementia. This suggests that depressive symptoms are common in AD, and that women with dementia not taking estrogen replacement may be prone to depression. So while estrogen therapy may not be beneficial for the cognitive symptoms of AD it may have beneficial effects on the mood disorders of the illness. It would appear that the studies that have failed to evaluate mood have not taken this beneficial effect of estrogen into account.

Estrogen: A Cure or a Prevention?

While estrogen cannot now be called a "cure" for AD, it may be a hormonal preventive, and if so, it could be a treatment with enormous, largely untapped potential. Before I explain the protective and therapeutic roles of estrogen in Alzheimer's dementia, I offer a brief overview of AD—its incidence, symptoms, diagnosis, and what are believed to be its primary risk factors.

ALZHEIMER'S: WHAT IT IS, WHOM IT STRIKES

Today, there are an estimated 4 million Americans with AD, and the number is projected to balloon to 14 million by mid-century if a cure is not found. Consider these salient facts about the incidence and impact of Alzheimer's in America:

- *Alzheimer's disease hits 600 out of every 100,000 people.*
- *AD is the leading cause of a loss of independent living and subsequent institutionalization.*[7]
- *The cost of this devastating illness approaches 100 billion dollars annually.*[8]
- *In men and women over eighty-five years of age, the prevalence of AD is approximately 50 percent.*[9]
- *According to the Alzheimer's Association, AD is the fourth leading cause of death in adults, after heart disease, cancer, and stroke.*

Here is one of the most revealing facts about AD: Two to three times as many women as men develop AD, and many neuroscientists now believe that estrogen loss plays a critical role in the genesis of this disease. The fear of developing Alzheimer's disease haunts countless numbers of menopausal patients who have memory lapses. Now there is hope that these fears may be eased by future research supporting the healing potential of estrogen.

Before detailing the estrogen story, an understanding of the disease and its risk factors is helpful. First, it's important to restate the obvious: AD is more common among the elderly. While the disease usually strikes people over age sixty-five, it can also affect individuals in their forties and fifties, though these are rare instances of "early-onset Alzheimer's." The link between AD and aging has been confirmed not only in the United States, but in Europe, Africa, and Asia. For instance, a study carried out in three rural villages in central Italy evaluated all forms of dementia in 968 inhabitants older than age sixty-four. The dementia rate was 1.1 per hundred between ages sixty-five to sixty-nine, increasing dramatically to 34.8 per hundred between ages ninety to ninety-six. Sixty-four percent of the dementia patients were diagnosed with AD and 27 percent with vascular dementia.[10]

Besides the increased incidence with age and female gender, what other risk factors have been identified? Here are the major factors:

- AD is more common among people with less education, and among those who are shown to be less mentally active. Neuroscientists theorize that the learning process stimulates the production of nerve cells, which means there is a larger reserve of these cells to protect against severe mental decline with aging and/or disease processes.
- A family history of dementia or AD itself, presumably based on inherited genes.[11]
- Specific genetic predispositions. People with late-onset AD are more likely to possess a gene on chromosome 19 called ApoE. Researchers at Boston University School of Medicine performed a statistical analysis of studies involving 5,930 AD patients and 8,760 controls. The researchers concluded that ApoE was a major risk factor for AD across all ages between forty and ninety years and in both men and women. The link between this gene and AD was present in all ethnic groups, although it was strongest among Caucasians. Overall, the ApoE gene abnormality is present in almost 90 percent of AD patients developing the disease late in life.[12]
- Individuals with Down's syndrome are highly susceptible to AD and

a threefold increase in the frequency of Down's syndrome is observed in the relatives of AD patients with a rate of 3.6 per 100,000 as compared to an expected rate of 1.3 per 100,000.

- A compromised blood flow to the cerebrum, for whatever reason(s), appears to be a cofactor in the development of AD.[13]
- Head trauma, aluminum in the drinking water, personal history of depression, diabetes, and high sugar consumption are other possible risk factors.[14]

Perhaps the most intriguing risk factor for AD is hormonal, and I say this because the hormone link may translate readily into promising new solutions for this devastating disease.

A growing number of reports point to estrogen deficiency as a risk factor for AD.[15] Thin women and women with a history of either heart attacks or hip fracture are more prone to develop AD.[16] Both of these conditions are caused, in part, by an estrogen deficiency state. In addition, women who've had a hysterectomy at an early age, either with or without removal of the ovaries, appear to develop a more severe form of Alzheimer's disease. The reason for this connection may be a reduced blood supply to the ovaries as a result of the surgery. Such studies suggest that estrogen replacement may prevent or ameliorate AD.

Beyond estrogen, several other factors seem to protect against AD. These include high occupational attainment, regular exercise, and the frequent use of anti-inflammatory drugs commonly used in the treatment of arthritis.[17] (One level of damage in the brain cells of AD patients involves inflammation, and nonsteroidal anti-inflammatory drugs reduce this inflammation.) Strangely, some studies suggested that smoking can be protective, and so can an increased incidence of headaches and hyperthyroidism in the form of Graves' disease.[18] (The evidence on smoking in no way suggests that people should risk dying from the three top killer diseases—heart disease, cancer, and stroke—on the off chance that they might lower their risk of the fourth most lethal disease.) A complete list of risks and protective factors is found in Table 5.

What happens to the brain in people with AD? A number of abnormalities have been identified, though neuroscientists are not fully certain whether they are causes or results of the disease. A sticky protein called beta amyloid has been found in plaques outside of neurons. One theory is that the AD patients have a genetic predisposition that leads to the formation of beta amyloid deposits. As the amyloid plaques increase, nerve cells become irretrievably damaged. This occurs as beta amyloid acts directly

TABLE 5 *Risk Factors for Alzheimer's Disease*

Confirmed Risk Factors
Old age
Family history of dementia
ApoE genotype (genetic factor)
Down's syndrome
Female sex
Other Possible Risk Factors
Ethnic groups
Head trauma
Aluminum in drinking water
Increased sugar consumption
Personal history of depression
Autoimmune thyroiditis
Diabetes mellitus
Alcohol abuse
Parkinson's disease
Sleep disturbances
Cerebral glucose metabolism defect
Underactivity
Possible Risk Cofactor
Compromised cerebral blood flow
Protective Factors
Estrogen replacement therapy
Anti-inflammatory drugs
High education level
High occupational attainment
Exercise
Headache
Smoking
Graves' disease

on blood vessels to produce excess oxygen radicals. These unstable molecules insult and disrupt vascular function, increasing the risk for early Alzheimer's disease. (The theory is comparable to the accepted notion that deposits of cholesterol build up in coronary arteries, causing atherosclerosis and subsequent heart attacks.) The oxygen radicals induced by beta amyloid yield high levels of a toxic by-product, hydrogen peroxide, which accumulates and causes nerve cell death and brain atrophy.[19]

The presence of beta amyloid plaques also appears to be associated with a reduction in the neurotransmitter acetylcholine, the brain chemical that mediates memory and many other basic cognitive functions. In a related development, scientists have identified an abnormal version of a protein called tau in Alzheimer's patients. The abnormal tau protein accumulates in the brain, and this appears to be yet another cause of acetylcholine deficiency. The loss of acetylcholine is also associated with nerve cell death and brain atrophy.[20]

Neuroscientists have also discovered tangled nerve cell fibers, called neurofibrillary tangles, in the brains of AD patients. These tangles, like the knots in stereo-headphone wires, are found in parts of the brain centrally involved in memory and cognition: the hippocampus and temporal and frontal lobes. The abnormal tau protein is found within these neurofibrillary tangles.

ESTROGEN AND ALZHEIMER'S DISEASE

In a 1997 article in the journal *Neurology* Dr. Victor Henderson stated that the evidence for a cause-and-effect relationship between estrogen therapy and the incidence of AD remains inconclusive.[21] However, numerous studies in women suggest that estrogen may block the development or progression of AD in a variety of ways. How do we know that estrogen protects against AD? There are both laboratory and clinical data supporting the preventative action of estrogen against Alzheimer's disease.

Animal models provide a convincing rationale for the role of estrogen in the treatment and prevention of AD. These studies establish the role of estrogen in the regeneration and preservation of nerve cells in the regions of the brain most frequently involved in the neurodegenerative changes associated with AD. Animal studies establish a relationship between the hormonal-dependent changes in brain anatomy that are involved in learning and memory. These laboratory studies provide insight on the estrogen and Alzheimer's connection, but the current excitement about estrogen for Alzheimer's disease stems from large observational studies and a few clinical studies of women taking estrogen. In toto, estrogen treatment in menopausal women has been shown to reduce the incidence of AD by 30 to 70 percent.[22] A risk estimate of 0.5 is a 50 percent reduction in the incidence of Alzheimer's disease. Both the highest dose of estrogen ($\geq$1.25 mg) and the longest duration

(≥15 years) reduced the incidence of Alzheimer's by approximately 40 to 60 percent. If the highest dosage is combined with the longest duration the reduction is approximately 80 percent. (See Figure 6.) Here is a partial summary of these "observational" studies:

- In a landmark 1994 publication, Dr. Victor W. Henderson, a neurologist at the University of Southern California, reported that estrogen therapy administered to menopausal women decreased the incidence of AD. In a follow-up study, Henderson and his colleague, Dr. Paganini-Hill, published the observation that estrogen administered in a variety of forms reduced the risk of AD. The Leisure World study included 8,877 women in a retirement community. Between 1981–92, 2,529 women died and 138 of these were identified as having AD. The risk of AD was reduced approxi-

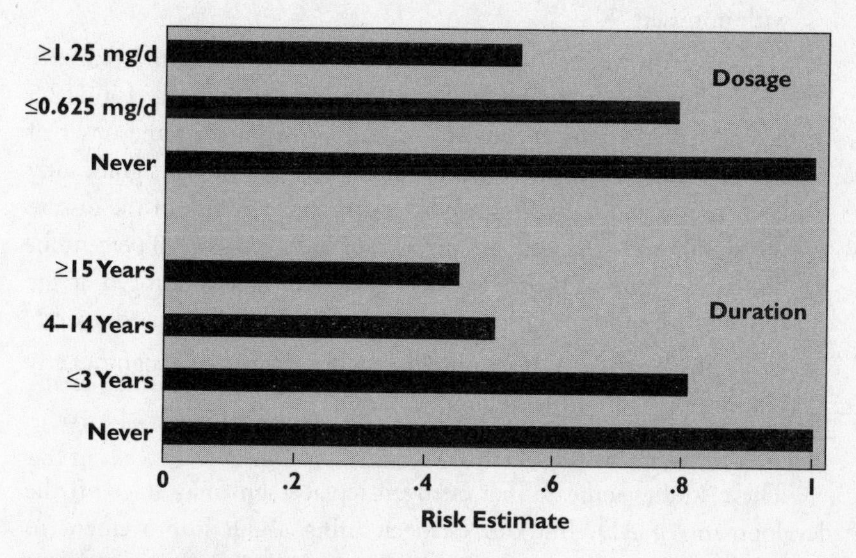

FIGURE 6 Estrogen Dosage and Duration and Alzheimer's Disease Risk

In the Leisure World retirement community cohort, there was a statistically significant relation between the relative risk for Alzheimer's disease and the dosage and duration of estrogen use after the menopause. These findings indicate that a higher estrogen dosage produced a 40 percent reduction in risk and the longest duration of estrogen use after menopause reduced risk 56 percent. Combining the highest dose with the longest duration produced approximately an 80 percent reduction. (*Data are from Paganini-Hill and Henderson, 1996. Permission from* Arch. Intern. Med. *1996; 156:2213–17.*)

mately 40 percent by the estrogen usage. Increasing estrogen doses and increasing duration of estrogen use both were related to the extent of protection against AD.[23]

- A study carried out at Johns Hopkins University followed 472 postmenopausal women for sixteen years. Thirty-four of these women developed AD during the follow-up. Approximately 40 percent of the women had used estrogen-replacement therapy. Comparing estrogen users with nonusers, there was an approximate 50 percent reduction in AD in the women who used estrogen.[24]
- In a 1998 Italian study, 2,816 women age sixty-five to eighty-four years were selected randomly from eight Italian municipalities. The women were divided into four groups based on age, each representing a five-year interval. Twelve percent of the women were on estrogen replacement for menopausal symptoms. These estrogen users had a 77 percent reduction in their incidence of AD, as compared with nonusers.[25]
- 1,124 elderly women who were initially free of Alzheimer's disease were followed over a period of one to five years. Overall, 156 (12.5 percent) of the women reported taking estrogen after the onset of menopause. The age and onset of Alzheimer's disease was significantly later in estrogen users compared to nonusers. The risk of the disease was significantly reduced, 5.8 percent for users versus 16.3 percent for nonusers. None of the 23 women who were taking estrogen at the time of their enrollment in the study developed Alzheimer's disease.[26]
- In a meta-analysis Yaffe reported that estrogen usage significantly reduced the risk for Alzheimer's disease.[27]

These studies indicate that estrogen replacement may stave off the development of AD. But can estrogen bring about improvement in patients who already have Alzheimer's? Four small clinical studies—two without control groups and two with control groups—explored this possibility. The two uncontrolled studies involved seven patients each, and both showed that estrogen brought about some reduction in the symptoms of dementia. The two controlled clinical studies, conducted at leading Japanese medical institutions, also produced positive results:

- Dr. T. Ohkura and his colleagues gave estrogen to fifteen women with AD and fifteen matched control subjects, and compared the results. Based on their measurements, the investigators concluded that estrogen replacement "significantly improves cognitive functions, dementia symptoms, regional cerebral blood flow, and EEG activity in female patients with dementia of the Alzheimer type." The only drawback: This was not a randomized, double-blind study.[28]

- Dr. H. Honjo and his colleagues conducted the only randomized, double-blind, placebo-controlled clinical study of estrogen for AD. Fourteen women with AD received either estrogen or a placebo. After six weeks, the estrogen-treated women compared to nonusers showed significantly greater improvement on a scale measuring symptoms of dementia.[29]

While these studies were indeed small, the positive results infer that estrogen is a viable treatment for women with Alzheimer's disease, and that many more large-scale clinical studies are needed to turn estrogen into a widely applied therapy.

How Estrogen Protects the Brain Against Alzheimer's

Fighting Beta Amyloid

As I've mentioned, AD is characterized by the accumulation of plaques consisting of beta amyloid that form in the brain, causing cell death and a disruption of normal communication between nerve cells. Physiological levels of estrogen reduce the generation of beta amyloid in cultures of embryonic human brain tissue. These findings suggest a mechanism by which estrogen treatment can delay or prevent AD. Another cell culture study showed that the higher the dose of estrogen, the less beta amyloid produced, which suggested a "dose-response relationship."[30]

Not only is beta amyloid directly toxic to brain cells, it increases the cells' vulnerability to other toxins, which increases the levels of toxic hydrogen peroxide and ultimately leads to cell death. Antioxidants such as vitamins C and E are thought to be part of a defense system against these toxic oxides, the by-products of so-called "oxidative stress." (This refers to the fact that certain oxidation processes produce cell-damaging oxygen radicals.) A recent German study found that the most powerful estrogen, 17-beta estradiol, protected brain cells from the oxidative stress that can destroy them. These data offer more support for the idea that estrogen should have a role in both the prevention and treatment of AD in postmenopausal women.[31]

Stimulating Nerve Growth Factors

Estrogen can also affect fetal brain development, with possible repercussions in old age. Estrogen stimulates so-called "neuronal growth factors" that support the functions of acetylcholine in the brain. (Again, acetylcholine is a primary messenger of cell-to-cell communication that forms the basis of memory and cognition.) Estrogen may act synergistically with these growth factors to favor the regeneration and survival of nerve cells, and to promote communication between them. This enhanced brain-cell communication, which develops early in life, is thought to protect an individual against the development of AD.

Restoring Acetylcholine

A deficiency of the neurotransmitter acetylcholine is a hallmark of AD. Low estrogen results in a shortfall of acetylcholine in the brain beyond what would normally be expected with aging. How does this occur? Estrogen deficiency causes an excess of an enzyme, acetylcholinesterase, that metabolizes—and inactivates—acetylcholine. Moreover, the normal growth of the brain can be hindered by a deficiency of estrogen. Brain cells responsive to acetylcholine require an adequate level of estrogen in order to be primed by the growth factors that stimulate their development.

Reviving Other Neurotransmitters

Though a shortage of acetylcholine is a central feature of AD, the disease is also associated with deficits in the three other key neurotransmitters, serotonin, norepinephrine, and dopamine. The reason? There may be too much of the enzyme monoamine oxidase (MAO), which degrades these neurotransmitters. In one study comparing AD patients with healthy controls of a similar age, the patients had higher levels of MAO in their brains and blood platelets, and the higher their MAO the worse their dementia.[32] Alzheimer's patients typically experience emotional deterioration, and those with more MAO had more abnormal psychological scores.

At the same time, certain genes seem to determine the extent to which mood-modulating neurotransmitters are deficient in AD patients. As such, MAO levels appear to be a genetic marker for the vulnerability of AD patients suffering from emotional problems. Drugs that inhibit MAO activity in the brain have been useful in the treatment of AD, as I will shortly elaborate.

Improving Blood Flow to the Brain

AD patients were also found to have an abnormally low cerebral blood flow. This reduced blood flow is not considered a sole cause but rather a "cofactor," which means that it contributes to the disease along with other risk factors. Experts agree that there is a high correlation between the disease and reduced blood flow—the extent of cognitive impairment parallels the decline in blood flow. In a Japanese study, sixteen patients with mild to moderately severe AD had cerebral blood flow evaluated by a sophisticated imaging technology, Single Photon Emission Computerized Tomography (SPECT).[33] Verbal IQ scores were low in patients with decreased cerebral blood flow of the left temporal and parietal lobes.

Working Together with Other Drugs

Estrogen can also enhance the AD patient's response to the Alzheimer's medication tacrine, marketed under the brand name Cognex. This drug inhibits an enzyme, acetylcholinesterase (AchE), that metabolizes and inactivates acetylcholine. The result? More acetylcholine in the brains of AD patients, which often yields improvements in memory and other cognitive functions. Another agent that functions in a similar way to tacrine is donezepil, brand-named Aricept. There is evidence to suggest that AchE inhibitors like tacrine also boost cerebral blood flow. These drugs, which usually have only modest benefits, work in ways that may complement estrogen. Researchers conducted a carefully controlled study in which 318 women were randomly put on either tacrine, a combination of tacrine and estrogen, or a placebo. After thirty weeks, the investigators found that patients taking the estrogen/tacrine combination performed significantly better on follow-up mental assessment tests than patients taking tacrine alone or a placebo. The patients on tacrine did show minor improvement on these tests, but the addition of estrogen enhanced the response in a robust fashion.[34]

The pivotal tacrine trial that led to FDA approval of this drug for treatment of AD also yielded information regarding the interaction of estrogen and tacrine. One hundred eighteen women with AD completed the trial. Thirty-seven were estrogen users and continued estrogen treatment concurrently with tacrine. When cognitive functioning was assessed, only the women receiving tacrine concurrently with estrogen had a significant beneficial response. (See Figure 7.) Tacrine alone did not produce significant improvement in cognitive function.[35]

I have found support for the use of such combination therapy in my own clinical practice.

HARRIET'S STORY: ESTROGEN AND ARICEPT

I am currently following the progress of a delightful seventy-five-year-old woman whom I've known for fifteen years. Harriet is just beginning

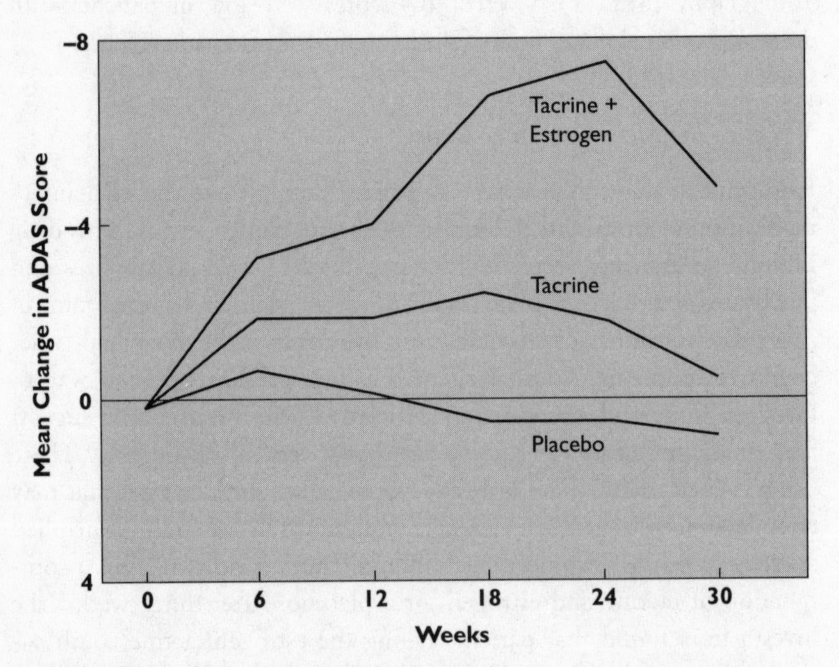

ADAS = Alzheimer's disease assessment score

FIGURE 7 Estrogen plus Tacrine and Alzheimer's Disease
Combined estrogen/Tacrine treatment for Alzheimer's disease produced significantly more improvement in mental status than Tacrine alone or placebo. (*Permission from Schneider, L. S., et al.* Neurology *1996; 46:1580.*)

to have serious symptoms of AD. Her psychotherapist, who suspected a hormonal factor in the depression and panic attacks she had endured for over a year, first referred her to me when she was sixty. Her menstrual periods had stopped at age fifty-five and she'd also been having a difficult time with hot flashes and night sweats. When I first saw her, she was taking a tricyclic antidepressant, as well as Xanax, an antianxiety medication, and had just completed a stress reduction program at a local hospital. It had been a hellish year for Harriet, whose appetite, which had always been hearty, had vanished; she was down twenty pounds to a body weight of eighty-eight pounds.

I was extremely concerned about Harriet's physical condition. She had smoked a pack of cigarettes daily for thirty years, and while I understood that it would be impossible for her to stop smoking when she was so emotionally upset, I hoped that she would respond to treatment and eventually wean off cigarettes. Hormonal studies revealed Harriet's low estrogen levels, typical of menopause, and I put her on a hormone-replacement regimen. Due to her depression and anxiety, I started her on a high dose of Premarin, 1.25 mg daily, and cycled her with a progestin. Her platelet MAO level was extremely high, and her psychiatrist felt that she might benefit by switching her antidepressant to an MAO inhibitor. On a combination of estrogen and an MAO inhibitor, Harriet gradually began to recover. Her appetite returned and her weight slowly rose until it reached a peak of one hundred and fifty-two pounds.

As Harriet's depression lifted and her anxiety eased, I told her my hope that she would stop smoking. She recognized the health risks involved, and she sincerely wanted to stop. "I'm willing to try," she told me, "but I know myself well enough, and if I become anxious I'll start again." Harriet was on target with her prediction: every time she tried to stop, her anxiety would surge and she'd resume smoking as a form of self-medication. But her body eventually would have the final say: At age seventy-three, her smoking-related lung disease worsened, and she had to be on oxygen twenty-four hours a day. Under these circumstances, Harriet was forced to quit smoking. Ironically, several years later she began to suffer some deficits in her mental alertness. (As I've noted earlier in this chapter, for unknown reasons smoking may confer some degree of protection against AD.) She began to have memory lapses, which were extremely upsetting to her. Harriet complained, "I get so upset, I want to use a word to tell my husband about something and it's just not there."

I continued to look forward to Harriet's monthly office visits, but I became increasingly concerned as I noticed changes in her response to my questions. I would often have to provide certain explanations repeatedly, which had never been necessary in the past. At first, I thought this resulted from a worsening of her breathing problem, but then I began to suspect—as did Harriet—that she was in the early stages of AD. It was reassuring when a Mini-Mental Exam showed good results, but clinically she complained of fatigue, worsening mental problems, and most alarmingly, increasing instances of falling. The latter is a frequent sign of declining central nervous system function. Harriet was also worried as she noticed a marked deterioration in her heretofore lovely and precise handwriting. I discussed her condition with several other doctors involved in her care, all of whom concurred, sadly, with the diagnosis of early AD. She began taking Aricept (donezepil), one of the two commonly prescribed AchE inhibitors I mentioned previously that are used to treat Alzheimer's. While concerned about this turn in her condition, I believed that her estrogen and MAO inhibitor would, together with the Aricept, protect her from a rapid progression of this insidious disease.

But my concerns were intensified by Harriet's subsequent office visits. She seemed to have more difficulty responding to my questions, and she said she'd fallen several times. Her husband was so concerned that he wouldn't let her walk on her own. I told her that I shared her husband's concern, encouraging her to be extremely careful and to consider using a cane or walker. "The only chance I get to walk alone these days," she told me, "is when my husband goes to the store without me. Then I get up and walk around my chair. It's the only time I can do it on my own." I believe that it's fortunate that Harriet started estrogen at age sixty, since it probably delayed the onset of AD. On the other hand, had she been treated even sooner—right after menopause—she might have benefited even more. Despite her gradually worsening condition, I still had hopes that she would not become completely incapacitated.

My hopes have not only been met but exceeded by Harriet's most current progress. It's been one year since she began the combination of Aricept and high-dose estrogen, and I have observed a striking improvement in her mental status. In our most recent visit, she was extremely alert and showed no signs of cognitive confusion. We talked openly about the diagnosis of Alzheimer's disease. "I was very worried I was going to get worse and worse," she said. "My husband already worries about me all the time, and if I have Alzheimer's I'll become a terrible burden on him." I reassured Harriet by telling her how impressed I was with her progress; she was clearly better than she had been one year earlier. "I still have some problems remembering the right word when I'm telling my husband something, but I do believe I am better," she said. "And I haven't fallen in quite a while." Of course, Harriet is but one patient. But her improvement and several of the research developments I have noted here all give me hope that in the near future we'll see improvements in the treatment of a disease that has long been viewed as hopeless.

Estrogen Alone for Alzheimer's Disease

A recent multicenter placebo-controlled study evaluated the effects of one year of estrogen treatment in women with mild to moderate Alzheimer's disease whose average age was seventy-five. At the end of the twelve-month study there was no difference between the estrogen and placebo groups. Apparently estrogen alone did not produce significant improvement in Alzheimer's symptoms. This study still leaves unresolved as to whether estrogen can prevent the development of Alzheimer's disease. Also unresolved is whether estrogen in combination with an acetylcholinesterase inhibitor, such as Aricept, can significantly improve symptoms of Alzheimer's disease as was suggested by a study we discussed earlier in this chapter.

ELIZABETH'S STORY: THE NEED FOR EARLY ESTROGEN

Think again about that striking statistic—50 percent of women over age eighty-five have some form of AD. Our mothers, grandmothers, aunts, or friends may be among these people, and we must reject the callous notion that quality of life is less important in octogenarians because their remaining life span happens to be short. And let's not forget about the millions of people in their sixties and seventies struggling with the disease, and the impact on their families.

While we wait for definitive medical solutions, we have sufficient evidence now to suggest that estrogen replacement has a potentially protective effect. Whenever estrogen is indicated for menopausal symptoms, heart disease prevention, and protection against osteoporosis, its anti-Alzheimer's effect should be considered as one more strong reason to commence with treatment—*especially* when any of the AD risk factors are present.

The story of one of my patients illustrates my point about early intervention with estrogen during menopause. Elizabeth was eighty-two when she first began to have memory problems. It was a particularly unfortunate development, since she had always been a bright, alert, independent woman. She had married at twenty-one, and lived a reasonably comfortable middle-class life. However, her life was not free of tragedy. Her first child, a daughter, died at age six months from a bacterial infection, and later, a son died at birth. After these tragic events, Elizabeth and her husband were told the devastating news that Elizabeth would not be able to have more children. Despite these predictions, when Elizabeth was thirty-five she gave birth to another son, and she later gave birth—at ages forty-two and forty-four—to another son and daughter, completing their family.

Elizabeth's last menstrual period occurred at age fifty-one, but she was not treated with hormone replacement until she was sixty. Her oldest son, a physician, had suggested that she take estrogen after she expressed concern about increasing facial hair. She remained in good mental and physical health until age seventy-five, when she contended with ulcerative colitis and an overactive thyroid. Surgery was necessary to remove part of her colon, and radioactive iodine reduced her thyroid function to normal. Shortly after her surgery, Elizabeth's beloved husband died. Despite her colostomy and the loss of her husband, she continued to live alone and take care of herself.

When Elizabeth's family threw a party celebrating her eightieth birthday, she still had a quick wit about her and enjoyed chatting with her old friends. Two years later, however, she began to have trouble with her memory. It was the beginning of AD, which she developed in spite of her overactive thyroid and her estrogen treatment, both thought to be protective factors against Alzheimer's. In my view, the failure of estrogen to completely prevent AD may have been due to the delay in starting treatment until a decade after her menopausal transition.

Elizabeth had another factor that increased her risk of AD. She had a definite sweet tooth. All her life she had prepared desserts such as chocolate and cherry pies and coconut cakes for her family, but the person who most enjoyed the fruits of her baking efforts was Elizabeth herself. During the last ten years of her life, the staple of her diet was sugar doughnuts. Studies suggest that a high sugar intake is associated with AD.

As the battle with AD began, Elizabeth had small lapses in memory and difficulty retrieving specific words. Gradually, this progressed to include other aspects of short-term memory loss. She began to repeat questions that had been answered ten minutes earlier. Her grandchildren complained to their mother that Grandma would repeatedly ask them their age and grade at school although they had answered those questions five times in the previous hour. She also became less capable of caring for herself. Her youngest son lived nearby and checked on his mother daily, which was a great consolation to the family, and a burden about which he never complained. Elizabeth could no longer prepare her own meals or reliably and properly take her medications.

Sadly, Elizabeth tripped on a rug and fell, breaking her hip, which required insertion of a pin to repair the fracture. Afterward, the family employed a young couple to live in her home. This could have made Elizabeth's life easier, but she became suspicious of the couple, convinced that they were stealing items from her house. No matter how strenuously her children reassured her about the young couple, she continued to grow more paranoid. One day, her wedding and engagement rings were missing and she was convinced that the couple had stolen them. Later, her children found the two rings where Elizabeth had hidden them, between the pages of an *Encyclopaedia Britannica* volume. Every week, Elizabeth's daughter drove one hundred miles to visit her mother and relieve the younger son of his heavy caregiving responsibilities. Eventually Elizabeth became so disabled that she could only sit briefly in a chair, spending most of her days in bed. The young couple could no longer care for Elizabeth, so she went to live with her daughter, and the family arranged for round-the-clock home care. She was very close to her daughter, son-in-law, and their three children and she felt surrounded by love.

Eventually, Elizabeth could no longer deal with the complications of her menstrual flow, and the family jointly decided to stop the estrogen that caused her to have periods. The cessation of estrogen seemed to mark a worsening of Elizabeth's condition. Her oldest son, Charles, who lived in another state, would visit once a month, but she no longer recognized him, pulling away when he bent over the bed to kiss her. He would have to sit in her room for a long time before some vague recollection reminded Elizabeth of who he was. "Charles," she'd call out, in the familiar voice she had used with him in the past, "please bring me a glass of water." Over the next several months, she became almost totally unresponsive, until she gradually drifted into a coma and died.

Elizabeth's story is typical of the devastation that can result from AD. And it poignantly emphasizes the need to develop early intervention to stave off this disease. The current information regarding the possible prevention of AD by estrogen therapy justifies the prompt initiation of estrogen treatment at the onset of menopause. While estrogen is not a surefire preventive, the best available research suggests that it can substantially reduce the risk and delay the onset of Alzheimer's disease, preserving the gift of consciousness.

The Pharmacologic Treatment of AD

As I mentioned earlier, the only FDA-approved medications on the market for the treatment of AD are two acetylcholinesterase (AchE) inhibitors, tacrine (brand name Cognex), and donepezil (brand name: Aricept). These medications inhibit the enzymatic breakdown of the neurotransmitter acetylcholine. This helps to reverse the acetylcholine deficit in the brain that is a hallmark of Alzheimer's. Other AchE inhibitors are currently being tested in clinical trials, and may soon be available. But the effects of these agents on the symptoms of AD are relatively modest. As I emphasized earlier in this chapter, a combination of tacrine and estrogen therapy was found to be more effective than tacrine alone in the treatment of patients with existing AD.

Oxygen radicals, by-products of normal oxidative processes in the body, can be toxic to nerve cells in individuals prone to AD. Thus, scientists have explored whether certain antioxidants have value in the treatment or prevention of this disease. One such antioxidant is vitamin E. One two-year-long study evaluated the efficacy of vitamin E, alone or in combination with an MAO inhibitor, selegiline, in the treatment of AD patients. The researchers tracked the progression of the disease in noninstitutionalized AD patients who were independent in at least two of the following activities: eating, grooming, and toileting. Patients receiving vitamin E (alpha tocopheral) or selegiline alone or in combination showed less progression of the disease than those receiving the placebo.

Only alpha tocopheral alone delayed nursing home admission. Adverse side effects were negligible and the cost was modest. In a number of controlled studies, the MAO inhibitor—selegiline—has compared favorably with other pharmacologic agents being used to treat AD. Selegiline-treated patients had an increased degree of independence in performing day-to-day activities, as compared to those receiving other drugs. Selegiline was well tolerated. The use of specific medications, including antidepressants, antipsychotics, and antianxiety drugs depends on the individual needs of the patients, and dosage must be carefully adjusted to reduce any side effects. Psychotherapies aimed at enhancing cognition are not demonstrably effective in AD patients, though they may still be useful for emotional symptoms in some patients.[36]

I have already documented a number of small clinical studies and trials of estrogen to improve mood and memory in Alzheimer's patients. I am pleased to report that two large-scale, well-controlled studies investigating estrogen for AD are currently underway under the auspices of the National Institutes of Health. One study is part of the Alzheimer's Disease Cooperative Study, and the other is sponsored by the Women's Health Initiative. The antioxidant properties of estrogen may be one of multiple mechanisms by which the hormone protects against beta amyloid and nerve cell damage. Estrogen also stimulates the growth of acetylcholine secreting nerve cells, helping to correct the cholinergic deficiency observed in all patients with AD.

The Future of Estrogen for AD

The dimensions of the medical, social, economic, and human problems stemming from Alzheimer's disease are staggering. An effective preventive or curative treatment would not only save the nation 80 billion dollars, it would alleviate untold suffering. While hormones alone may not be the cure, estrogen now appears as a promising treatment that may partially stem the tide of this disease. The use of estrogen in the prevention of Alzheimer's appears very promising. Without estrogen prevention 50 percent of women over age eighty-five will have AD, and if they live longer the percentages become even higher. With estrogen use of over a one-year duration the reduction in the incidence of AD was approximately 80 percent. (See Figure 8.) Researchers should help us understand how best to use estrogen—alone or in combination with other drugs or hormones—to maximal benefit for patients. Should the value of hormone replacement be confirmed, everyone will benefit: patients, their families, doctors, and the health care system as a while. The cost of a year's supply of HRT for women is one fifth the amount it takes to care for an Alzheimer's patient for that same year.

While the evidence regarding estrogen and AD is hopeful, we still need those large-scale studies before doctors can say with utmost certainty that the hormone decreases the risk or severity of Alzheimer's. This means following thousands of women for many years in a research design that places some women on estrogen and others on placebo. Fortunately, the Women's Health Initiative Memory Study may fill the bill. It will determine whether HRT reduces the risk of AD in thousands of postmenopausal women being followed for ten years. We already know that estrogen users are better educated and economically more secure than nonusers. These factors, independent of estrogen, may influence whether people develop AD. But the Women's Health Initiative Memory Study

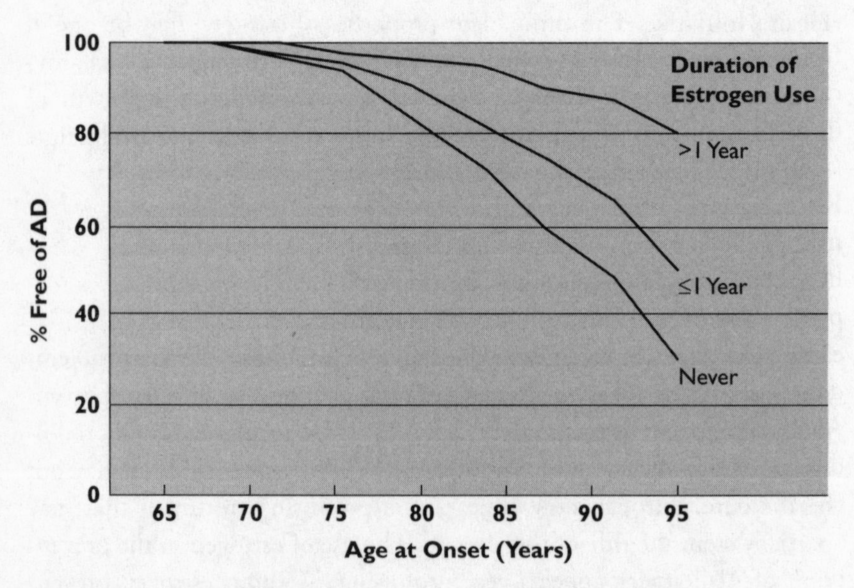

FIGURE 8 Estrogen Delays Onset of Alzheimer's Disease
A study of 1,124 menopausal women who were free of AD at baseline were followed for fifteen years. The curves in figure 8 show the percentage of population that is free of disease at certain ages. Women who took estrogen for more than one year experienced a dramatic delay in AD onset. (*Permission from Tang, M. X., et al.* Lancet, *1996.*)

should tease out which factors truly prevent Alzheimer's, including estrogen use. While I am heartened by the wealth of research confirming the positive effects of estrogen on the brain, we still need that hard data from large studies like the WHI. When the numbers are in, we may have a far more effective set of weapons in the fight against this devastating illness. Until that time, women concerned about AD should consider all the available evidence, including the entire range of benefits and risks, as they decide whether to commence with hormone modulation.

The actual definitive diagnosis of AD can only be made following an autopsy, when the brain can be examined under the microscope. The German neurologist Alois Alzheimer (1864–1915), describing a patient, made the original report of Alzheimer's disease. An autopsy examination of the brain revealed degenerative changes in the nerves. Plaques clogged nerve function and tangled the nerve cells, preventing normal neurotransmission. The plaques consisted of beta amyloid.

For neurologists or other physicians to diagnose Alzheimer's disease, they must first rule out other conditions with dementia or dementia-like symptoms, such as hypothyroidism, B_{12} deficiency, kidney or liver malfunctions, pernicious anemia, or pellagra. Blood tests, neuropsychological exams, and mental status questionnaires can help clinicians to make distinctions between Alzheimer's dementia and other conditions.

A number of sophisticated imaging technologies, including Magnetic Resonance Imaging (MRI), Single Photon Emission Computerized Tomography (SPECT), and Positron Emission Tomography (PET) scans have been used to identify brain atrophy, which is a diagnostic criteria for patients with AD. These modern techniques may be useful in spotting early AD. (A 1997 study from the Mayo Clinic using MRI techniques described atrophy of the temporal lobe of the brain as an early sign of AD.) It is now known that the mental impairment of an AD patient is directly related to the degree of brain atrophy, so high-tech imaging can be used to track the progression of the disease.

The signs and symptoms of AD are wide ranging and involve multiple functions of the central nervous system. I've listed some of the major symptoms below. (See Table 6.) The disease begins with a loss of memory, primarily for recent events, accompanied by difficulty retrieving words. An early sufferer of AD may meet someone new, but the next day when he or she sees the person again, it's as if they'd never met. The disease can then progress to an inability to perform one's job or to carry out the mundane activities of daily living. A hotel maid is fired because she can't remember what room she has just cleaned, or a grandmother no longer can bake cookies because she doesn't remember to turn on the oven.

Falling is a frequent problem among the elderly, but it is far more common and severe among people with AD. The brain is a critical factor in maintaining equilibrium, and deterioration in the brain, such as occurs with AD, can wreak havoc with balance. One of the brain processes involved in maintaining equilibrium is called proprioception, which is the unconscious perception of movement and spatial orientation; it tells you precisely where your feet are without looking. It is what allows you to walk up the attic stairs carrying a box without looking at your feet. One of the changes that occurs with aging is the way you fall. Younger people tend to fall forward, curling up as they fall to allow them to roll onto the ground in a ball, protecting themselves against major injury. Elderly people, whose aging brains no longer can control the reflexes that allow them to fall in a protective manner, usually fall backward. The portion of the body that strikes the ground first at full force is the hipbone. If the bone has been weakened by osteoporosis, the

TABLE 6 *Symptoms of Alzheimer's Disease*

Relentless impairment of memory and reasoning, beginning insidiously and progressing for a decade or longer
Failure to learn new recent events
Defect in recognition of previously known places or situations
Inability to retain employment
Unable to cope with activities of daily living
Lack of interest
Unwise decisions regarding property and investments
Poor judgment
Deterioration of social relationships
Paranoia during the evening hours
Nighttime disorientation
Psychotic thinking
Impairment of language
Deficit memory for words
Inability to name specific objects and people
Compromising visual attention (looking at one thing at a time)
Loss of smell

force of the fall will almost certainly result in a fracture. Ten billion health care dollars are expended each year to care for people sixty-five or older who are injured in falls. (The enormity of this problem underscores the need for preventive treatments. In this regard, the use of estrogen in menopausal women can have a dual action: reducing risks of the Alzheimer's dementia that leads to more falls, and preventing the thinning of bones due to osteoporosis that makes fractures more likely.)

As the disease progresses, the individual's unique personality may seem to fade. Judgment is impaired and relationships with family may become difficult; nighttime hours can be particularly hard, with outbursts of paranoia and disorientation. The capacity for speech may be lost, and some patients eventually lose consciousness and die from this horrific disease.

7

The Protected Passage

HORMONE MODULATION FOR A HEALTHY MENOPAUSE

Today's women no longer take their physicians' word as law. They wish to be informed by their doctors of health care choices, then come to their own decisions. I applaud this transformation. All women have the right to expect their medical treatments to be tailored to their specific needs. This new mind-set is most apparent when women confront their choices of treatment for menopausal symptoms. The approach I advocate—hormone modulation—treats each woman as a unique biological and psychological individual.

Individualization of medical treatment for any condition is always important, but when it comes to menopause, it is imperative. Once hormone therapy has begun, doctor and patient must closely monitor her response to treatment. Hormonal blood levels should be tested, and treatment should be appropriately altered as needed. Unfortunately, too often this is not the approach being taken today with standard hormone replacement therapy, or HRT. Today's HRT frequently consists of a fixed dose of estrogen and progesterone for each and every patient. The standard dose does not account for individual variations in patients' needs.

I have used the term "hormone modulation," to refer to an individualized form of hormone treatment, which may include estrogen, progesterone, testosterone, and in some cases thyroid hormone in doses tailored to the individual patient. I use this term to emphasize the individualization and flexibility of therapy. Namely, the types and doses of hormones are adjusted to most effectively and safely relieve an individual woman's symptoms. I strongly feel that women undergoing hormone treatment should have their hormonal blood levels checked regularly to be sure they remain in a range that is both safe and effective.

Why is individualization so important? Recent studies have shown that the exact same dose of estrogen given to different women can result in blood levels that vary by as much as tenfold.[1] Researchers have seen this tenfold variability regardless of how the hormones are given (oral, patch, etc.) The reasons for these variations are not always clear, but they probably relate to the fact that each woman absorbs and metabolizes estrogen differently. My own clinical experience has also taught me that during a course of treatment, one woman's estrogen levels can vary even as she stays on a fixed dose, so she may need adjustments in her medication every few months or years.

Hormone modulation is based on the idea that people respond uniquely to medications of almost every type—especially to natural biological substances, e.g., hormones, or drugs that mimic those substances. A recent *New York Times* article cited the work of Dr. Jay S. Cohen, an associate professor of psychiatry at the University of California at San Diego, who has criticized the doctor's pharmaceutical bible—The Physician's Desk Reference (PDR)—for offering "one size fits all" dosing recommendations. While Dr. Cohen's main focus is dosing recommendations for various drugs that are too high, leading to adverse side effects, he emphasizes the need to individualize doses of drugs to get both the safest and the most effective regimens. I agree that dosing recommendations for many drugs are too high, but the opposite can also be true. When it comes to hormone therapy, I've seen women who've been put on doses of estrogen or progestin that are either too high or too low.

"People vary greatly in their sensitivity to drugs, [Dr. Cohen] said, and one person's remedy may be another's overdose," writes Denise Grady in the *Times* article. "Although weight, age and sex help determine how a dose of medicine will affect someone, it is also possible for two people who are identical in those traits to react quite differently to the same dose of the same drug." I have found this to be absolutely the case in my treatment of peri- and postmenopausal women. I also agree with comments in Grady's article attributed to Dr. Peter Honig, deputy director of the office of postmarketing risk assessment at the Food and Drug Administration. "We learn a lot about a drug after it's been approved," said Honig. "I think we would agree with that. I think [Dr. Cohen] makes some good points. Individualized dosing is the way to go. But it's not well taught in medical school."

Not only do I concur with Drs. Cohen and Honig, I believe that this profound principle of biological individuality must be taken farther. Namely, it not only matters what *dose* women take of a hormone, it matters *what type of preparation* they take. There are a large variety of hormonal preparations that can be used in hormone replacement therapy for menopause. Given this diversity, questions have arisen about which types of hormones are best for preventing which menopause-associated diseases, such as heart disease, osteoporosis, and Alzheimer's disease. Research currently is under way to answer some of these questions, but much more is needed.

We must still individualize treatment, because generalizations based even on the best clinical research may still not apply to everyone. In this book, I have focused on the influence of hormones on the mind, and my treatment decisions made with menopausal patients often depend upon their potential effects on mood, memory, and sexuality. Certainly, though, women and their doctors must also consider other critical health issues as they make treatment decisions.

To date, most physicians have focused exclusively on how HRT influences the heart, bones, breast, uterus, and physical symptoms such as hot flashes. I would add the brain to this list, with all the potential symptoms of a hormone imbalance in the brain—depression, memory loss, cognitive confusion, dementia, PMS, or sexual dysfunction. This whole new dimension of consideration for treatment of the whole woman means that doctor and patient must be that much more sensitive to biological individuality.

BEYOND STANDARD DOSES OF HRT

If standard doses of HRT were the best available treatment for *all* menopausal women, we would not see the problems with mood, memory, cognition, and sexuality that are all too prevalent. There is no question that standard HRT works well for some women, but for others it is a prescription for trouble. Treatment must be custom-fitted to each patient's physical symptoms and her long-term health goals. If her goals include mood stabilization and/or mental sharpness, standard-dose HRT is going to be particularly tricky—either inadequate or outright deleterious.

Once a woman has made the decision to start HRT, standard treatment consists of an estrogen and progestin. Estrogens are taken either orally on a daily basis or once or twice weekly as a transdermal patch. An oral progestin is administered either daily or cyclically for ten to fourteen days each month. Daily progestin frequently eliminates a menstrual flow because of its suppressive effect on the lining of the uterus (endometrium). Cyclical progestin induces a menstrual flow that occurs when the progestin is stopped. Both of these progestin regimens are effective in protecting against uterine cancer. (A progestin need not be administered if there is no uterus.) Based on my clinical experience, research, and other investigators' studies, I am certain that women respond differently based on the dose and type of estrogen. Also important, some women respond very negatively to certain types, doses, and schedules of progesterone. I will delve into the progesterone question shortly.

Fears and concerns about breast cancer make many women reluctant to start HRT. But how many women are told that only 3.8 percent of American women who live to ninety will get breast cancer, while 50 percent will die of heart disease? Patients are often told that HRT cuts the risk of heart disease by as much as half, but few realize how much more prevalent and deadly heart disease is for women, as contrasted with breast cancer.

Physicians must spend considerable time with their patients to fully clarify the risks and benefits of hormone treatment. Unfortunately, doctors are increasingly hemmed in by the time constraints imposed by managed care and HMOs. These constraints erode their ability to educate their patients regarding the pros and cons of this critical decision. Too many menopausal women see their doctors for one brief visit, are given a prescription, then told to return in six months. Studies show that a majority of these women never fill their prescriptions, or quickly discontinue treatment because they do not feel that they are being adequately followed. Many women I have seen are both aware and concerned about the lack of individualization of their treatment. They fear they are taking medicine without fully understanding its effects on their whole body. Menopausal women who are not having hot flashes or night sweats may not notice any immediate benefits from estrogen treatment, and they frequently stop taking their medication without realizing that the protective health benefits on the heart, bones, and brain require long-term therapy.

In 1997 a study from Emory University School of Medicine in Atlanta, Georgia, found that postmenopausal hormone replacement therapy by women physicians in the United States was higher than reported in cross-sectional U.S. surveys of nonphysician women. Approximately 60 percent of women physicians aged forty to forty-nine used HRT and these female physicians were more likely to be gynecologists.[2] This may be predictive of greater use of HRT for U.S. women in the future.

As more and more women use HRT they should be concerned about the long-term effects of hormone treatment, both negative and positive. To make informed decisions about their present and future health, they must have all the information—not just about breast cancer, but also about heart health, bone health, and brain health.

Hormonal treatment impacts all organs of the body, including the brain. Just as the absence of hormone therapy can negatively impact the mind, the wrong hormone therapy can also negatively impact the mind. I've seen numerous women suffering from depression, insomnia, and lethargy while taking various forms of hormone treatment, including oral contraceptives. Their marriages are in turmoil because they have no interest in sex. When I carefully modulate their hormone therapy, their symptoms are relieved. Such patients often say, "Why didn't somebody tell me my hormone treatment could cause these problems?" It is a very difficult question to answer. Many physicians are not fully aware of the multileveled impact of hormones on the brain. Doctors understandably have trouble keeping up with all the latest medical advances. That is why we specialize, but this has been both our blessing and our curse. As we physicians specialize more and more, we become increasingly isolated from one another. Each of us is trapped in our own narrow box of biomedical interests.

One of the major reasons I decided to write this book is to create a better awareness on the part of patients and physicians for the need of a cross-disciplinary approach to medical care. This is no theoretical argument; it's about effective patient care in an era of managed medicine. When it comes to menopause, primary care doctors, gynecologists, endocrinologists, neurologists, psychiatrists, and psychologists must develop an individualized treatment approach that enables women to protect the health of their bones, breasts, hearts, brains, and reproductive organs—not just one or two of these body parts. And once the brain enters the equation, a woman's right to and desire for a healthy mind as well as a healthy body will be properly regarded by the medical profession.

HORMONE MODULATION: VARIETIES OF ESTROGEN AND PROGESTERONE

Types of Estrogen

Menopausal women frequently ask whether I prefer one type of estrogen over another. I have no specific estrogen that I prefer over another. The type of estrogen prescribed is a decision that is jointly made by the patient and myself. Some women have very definite ideas as to the types of estrogen they will or will not take. The estrogen preparation that has been on the market the longest is Premarin, which has been available for over fifty years. Most all the early studies demonstrating the positive effects of estrogen on bone and heart and more currently on brain have been done using Premarin. The name Premarin is derived from the source of the estrogen, *pre*gnant *mar*es ur*in*e. From this source Premarin is extracted for use as an estrogen replacement. The use of pregnant horses to supply the urine has become controversial. Animal lovers and horse lovers in particular feel that this is an exploitation of horses. My understanding is that these horses are treated extremely well and I myself have no ethical problem in prescribing Premarin. In fact, because we have fifty years of experience to call upon in assessing its efficacy and safety, I feel very comfortable using this form of estrogen. Premarin also offers a greater number of estrogen doses (0.3, 0.625, 0.9, 1.25, and 2.5 mg) than do other estrogen preparations, since it is essential to individualize dosage to meet the needs of each woman. Premarin contains primarily estrone sulfate, which when administered orally is absorbed from the gut and passes through the liver before entering the rest of the body. This "first pass" through the liver may have an advantage in that it allows estrogen to stimulate the production of HDL, or good cholesterol, by the liver, more than would occur if estrogen were administered through the skin (transdermally) where no liver "first pass" would occur. Premarin also contains some animal estrogen, in particular, significant quantities of equilin, which has biologic activity, particularly on the brain. Animal studies done at the University of Southern California in Los Angeles have demonstrated that equilin induced increases in nerve cell growth in the cerebral cortex.[3] Because Premarin has been available for over fifty years, for many physicians it is the estrogen of choice for HRT. They feel comfortable because fifty years of experience would have brought to life any major safety issues and because its efficacy over the years has clearly been established.

Having said this, however, I am very comfortable prescribing other estrogens. Estradiol, which is found in all current transdermal preparations and in Estrace, a brand of oral estradiol, is an extremely effective estrogen and has become popular with menopausal women because they consider it to be a "natural" form of estrogen. The reason women call estradiol "natural" is because estradiol is the estrogen that circulates in the blood of the human female. Estradiol has the strongest estrogenic activity of any estrogen on the multiple organ systems, which respond to estrogen treatment. Estrone, the major estrogen found in Premarin, is converted by the liver to estradiol. Currently there is a somewhat popular "fad" in which women wishing to use natural estrogen are using a combination of estriol (an estrogen of pregnancy), estrone, and estradiol, which some women feel is a more natural way to take estrogen. This is a serious mistake. Estriol has very little estrogenic activity and the amounts of estrone and estradiol in these preparations have been so low that they do not exert a therapeutic effect.

The types of estrogen available to treat menopause are listed in Table 7. These include oral, transdermal (patch), and vaginal preparations. The vaginal preparations are used primarily to eliminate vaginal dryness and atrophy. Because there is less systemic absorption of the estrogen, vaginal preparations do not relieve hot flashes or night sweats nor do they have the health benefits of oral and transdermal estrogens.

Types of Progesterone

Oral, vaginal, and transdermal progestin preparations are listed in Table 8. Shortly, I will explore in detail the rationale and controversies regarding the use of progesterone in the hormonal treatment of menopause. Put simply, progesterone must be given in some form because estrogen alone—so-called "unopposed estrogen"—has been linked to an increased risk of uterine cancer. Progesterone protects the uterine lining from abnormal growth and the development of cancer. As far as we know,

TABLE 7 *Types of Estrogen and Estrogen/Progestin Combinations*

ORAL PREPARATIONS		
Trade Name	**Generic Name**	**Dosage**
Premarin	Conjugated equine estrogen (estrone sulfate and equilin sulfate)	0.3–1.25 mg daily
Cenestin	Synthetic conjugated estrogens	0.625–0.9 mg daily
Ogen	Estropipate	0.625–2.5 mg daily
Ortho-Est	Estropipate	0.625–1.25 mg daily
Estrace	17-Beta-estradiol	0.5–2.0 mg daily
Estratab	Esterified estrogen	0.3–2.5 mg daily
Menest	Esterified estrogen	0.3–2.5 mg daily
Combined Estrogen/Progestin		
Prempro	Conjugated equine estrogen medroxyprogesterone acetate	0.625 mg/2.5 mg daily
Premphase	Conjugated equine estrogen/medroxyprogesterone acetate	0.625 mg—days 1–14 0.625 mg/5 mg—days 15–28
Ortho-Prefest	Estradiol/norgestimate	1 mg Estradiol for 3 days followed by 1 mg Estradiol/Nongestimate (0.09 mg) for 3 days, repeatedly
TRANSDERMAL PREPARATIONS		
Estraderm	17-Beta-estradiol	0.05–0.1 mg, twice weekly
Climara	17-Beta-estradiol	0.05–0.1 mg, once weekly
Vivelle	17-Beta-estradiol	0.0375–0.1 mg, twice weekly
Alora	17-Beta-estradiol	0.05–0.1 mg, twice weekly
Fempatch	17-Beta-estradiol	0.025 mg, once weekly
Combined Estrogen/Progestin		
CombiPatch	17-Beta-estradiol/norethindrone acetate	0.05 mg/0.14 mg twice weekly
VAGINAL PREPARATIONS		
Estrace Vaginal Cream 0.01%	Estradiol vaginal cream 0.01%	2–4 g daily for 2 weeks Maintenance: 1 g 3 times a week
Estring Vaginal Ring	Estradiol vaginal ring	2 mg for 90 days

TABLE 7 *Types of Estrogen and Estrogen/Progestin Combinations*

ORAL PREPARATIONS

Trade Name	Generic Name	Dosage
VAGINAL PREPARATIONS		
Ortho Dienestrol Cream	Synthetic, nonsteroidal estrogen	1–2 applications daily for 2 weeks Maintenance: 1 application 3 times weekly
Premarin Vaginal Cream (0.65 mg/g cream)	Conjugated equine estrogen (estrone sulfate and equilin sulfate)	0.5–2g daily

there is no other compelling reason to give progesterone, but of course, preventing uterine cancer is a very compelling reason.

The progesterone in standard HRT is given in the form of synthetic progestins, such as medroxyprogesterone (Provera) and norethindrone. Generally, these progestins are given as a pill taken orally, either on a daily basis or cyclically—10 to 14 days each cycle.

Synthetic oral progestins cause many women to suffer physical side effects, such as bloating, breast tenderness, fluid retention, hot flashes, and night sweats. The combination of estrogen and synthetic progestins may produce a PMS-like state including anxiety, depression, irritability, insomnia, confusion, and decreased sexual drive and function. The reason, as I have explained in previous chapters, is that progestins have antiestrogenic activity. Norethindrone specifically has chemical similarities to testosterone, which also has antiestrogenic properties.

One interesting approach to postmenopausal hormone replacement is Ortho-Prefest containing estradiol and a synthetic progestin, norgestimate. Estradiol alone is taken on days 1 to 3 of the therapy and the combination estradiol/norgestimate treatment is taken on days 4 to 6. The pulsative pattern is repeated continuously to produce a constant estrogen/intermittent progestin regimen. This novel approach produces fewer negative side effects from the progestin.[4]

My research has shown that some progestins appear to alter the metabolism of estrogen, slowing its elimination from the body and raising estrogen blood levels.[5] I am concerned that these high estrogen blood levels may put women at an increased risk of breast cancer.

TABLE 8 *Types of Progestins*

Trade Name	Generic Name	Daily Dosage
ORAL ADMINISTRATION		
Amen	Medroxyprogesterone	5–10 mg daily, 10–14 days
Aygestin	Norethindrone acetate	2.5–10 mg daily, 10–14 days
Cycrin	Medroxyprogesterone	2.5–10 mg daily, 10–14 days
Provera	Medroxyprogesterone	2.5–10 mg daily, 10–14 days
Prometrium	Micronized progesterone	200–400 mg daily, 10–14 days
VAGINAL PREPARATIONS		
Compounded micronized progesterone vaginal suppositories	Micronized progesterone	100 mg morning and evening
TRANSDERMAL PREPARATIONS		
CombiPatch	17-beta-estradiol/ Norethindrone acetate	0.05 mg 0.14 mg Twice weekly

(When I prescribe somewhat higher-than-usual dosages of estrogen, I do so to raise a woman's low blood levels to normal levels, not to abnormally high levels. Progestins may be responsible, in some cases, for pushing a woman's blood estrogen to an unacceptably high level.) Recently there have been studies that suggest that the progesterone in HRT increases the risk of breast cancer compared to estrogen alone. I will discuss this issue in more detail in chapter 9.

Micronized progesterone is synthetically produced from chemicals that are extracted from the Mexican yam. Progesterone is poorly absorbed in its natural form through the intestinal tract. Micronization is a process by which the synthesized progesterone is ground to extremely fine particles (micronized), which can be more easily absorbed through the gastrointestinal tract. Micronized progesterone is now available in an FDA-approved capsule called Prometrium, available through all pharmacies. However, vaginal micronized progesterone must be compounded by a licensed compounding pharmacist who on doctor's orders will prepare whatever dosage the physician wishes to prescribe. Most physicians in large cities would be aware of the compounding pharmacist in their area. A compounding pharmacist also prepares methyltestosterone in the low doses I prescribe.

Vaginal progesterone, given in the form of a suppository, is a good alternative for one reason. Vaginal suppositories appear to deliver natural progesterone more directly to the uterus, where it is needed to prevent uterine overgrowth and malignancy. The theory behind the use of the vaginal suppository is that it will result in less systemic absorption, and, therefore, causes few side effects in the rest of the body—including the brain. This is the theory, but the theory most certainly needs to be proved by clinical research that will test its validity—more on this in a moment. I have obtained blood progesterone levels on women using vaginal progesterone suppositories and there is definitely some measurable increase in blood progesterone levels. However, in the clinical experience of other physicians, as well as in my own experience, the use of vaginal progesterone frequently reduces the negative mood and cognitive effects associated with oral progestins. In many of my clinical cases, the switch from a synthetic or even a natural progesterone given by mouth to a vaginal suppository results in a dramatic improvement in the patient's psychological state. Some women object to the vaginal progesterone preparation because it is messy. It can require the use of a pad for several hours after insertion. But many patients gladly put up with this inconvenience in exchange for fewer emotional and physical side effects. Again, each woman's therapy must be tailored to her individual needs and comfort level.

That said, I must emphasize that no well-controlled, comparative studies have evaluated the differences between synthetic progestins such as medroxyprogesterone and norethindrone versus micronized progesterone. We also need studies that compare the vaginal versus the oral route of administration of micronized progesterone. I base my recommendations on the extensive experience of other menopause experts and my own large clinical experience. This method of progesterone administration has not been approved by the FDA and has not been evaluated as extensively as the oral synthetic progestins and oral micronized progesterone. However, when I am confronted with a woman who has serious adverse reactions to those forms of HRT I am faced with a quandary. Should I take her off all forms of progesterone treatment and monitor carefully with endometrial ultrasounds and biopsies? This approach to the problem has been recommended by a group of menopause experts as the way to treat such patients and was in an article published in 1989 in *Maturitas* entitled, "Consensus statement on progestin use in postmenopausal women."[6] The authors recognize that this approach increases a woman's risk of developing uterine cancer. I avoid, if at all possible, taking such an increased risk. That is why I have chosen the vaginal progesterone regimen, which in my opinion minimizes adverse reactions, yet provides adequate protection against uterine cancer.

PROGESTERONE: PROBLEMS AND CONCERNS

MARCY'S STORY

At age forty-two, Marcy had just started a new business. It was a stressful endeavor, and she began to have trouble with insomnia and a declining interest in sex. She assumed that both these symptoms were stress-related. When she began waking up at night dripping with sweat, she realized that something more than stress was involved. Her doctors ran blood tests, which revealed a sky-high FSH level—a sure sign of early menopause. She began daily treatment with a combination of estrogen and progestin.

Almost immediately, Marcy began to feel bloated, and her breasts became extremely tender. "I was so uncomfortable, I couldn't even sleep on my stomach," she later said. Her doctor cut the dose of the estrogen in half and the physical side effects subsided, but the insomnia returned and her sex drive became even flatter. Her physician then referred her to me for further hormone evaluation. Marcy's blood estradiol and total and free testosterone levels were all quite low. After hearing her story, my first impression was that she was extremely sensitive to her daily progestin. This sensitivity occurs in many women, often prompting them to discontinue HRT. I recommended that she stop the daily progestin, resume her estrogen, and start methyltestosterone to revive her libido. I prescribed cyclic micronized progesterone, administered the first ten days of each calendar month, in order to produce regular menstrual cycles. I hoped and believed that this regimen would relieve her insomnia and restore her sex drive.

When I saw Marcy six weeks later, she was feeling much improved. "My job had nothing to do with my symptoms. I am still working just as hard as ever—or even harder—and I feel fine now. Not taking that progestin every day has turned my life around. My sex drive is back and I'm sleeping eight hours a night. My breasts are not tender and I can go back to sleeping on my stomach."

Marcy's extreme sensitivity to daily progestin is not very unusual. Ten years ago, my colleagues and I received support from the NIH to evaluate the extent of progestin sensitivity in menopausal women. Our goal was to determine factors that might predispose a woman to be so sensitive to her progestin that she would eventually stop taking estrogen to avoid the horrific side effects of the medication. Our suspicion was that many women with progestin sensitivity would have a history of mood difficulties, and that their brain chemistry would predispose them to depression and anxiety.

We would follow women on HRT and check their estrogen and testosterone levels before and during hormone replacement. (We did not include women with a history of major depression, since we wanted results that were applicable to the broad population of menopausal women.) The HRT consisted of daily estrogen with the addition of progestin for ten days each month. We obtained mood and hormone measurements when the women were taking estrogen alone, and during the ten days when they took both estrogen and progestins. We discovered that nearly two thirds—65 percent—of the women exhibited progestin sensitivity.[7]

Who were these progestin-sensitive women? They tended to have the lowest estrogen and testosterone levels before HRT began. Compared to women who were not progestin-sensitive, these women had a more extensive history of mood disorders, and a lower platelet MAO level before beginning treatment. The lower MAO implied that abnormalities in brain neurotransmitter function contributed to their progestin sensitivity. (In previous chapters, I have emphasized the relationship between abnormally high MAO and mood disorders, but evidence also exists that abnormally low MAO is associated with neurotransmitter malfunctions that affect mood.)

The progestin-sensitive women also had a longer duration of menopause (more than two years) before starting HRT, whereas women not so sensitive to progestins had started treatment within the first two years of their menopause. The sensitive women with a longer duration of menopause had blood estrogen and testosterone levels that were roughly half those of the women with a short menopausal duration.[8] Their relative estrogen deficiency partially explained their abnormal MAO activity and mood disorders. Dr. Barbara Sherwin has made a somewhat similar observation in a study showing that menopausal women on a high dose of estrogen had fewer mood disturbances when taking a progestin than women on a lower dose of estrogen.[9]

Our research suggests parameters that you and your doctors can use to predict whether you are likely to have a negative reaction to synthetic progestin. Have you been menopausal for two years or longer? Are your estrogen and testosterone levels already quite low? Have you had a history of mood swings or depression? If so, you may find natural micronized progesterone an acceptable alternative to synthetic progestins.

The Introduction of Progestins into HRT

To put the progesterone question in proper context, a bit of medical history is useful. In the 1960s and 1970s, menopausal women were often given estrogen without any form of progesterone. Some physicians advised their patients to stop estrogen therapy for one week each month. This regimen sometimes resulted in a menstrual flow. Understandably, the continuation of a menstrual flow following menopause was unpleasant for most women, and many women urged their physicians to administer estrogen replacement in a way that would abolish periods. Physicians complied and prescribed continuous estrogen, thus eliminating periods. The use of so-called "unopposed estrogen" (estrogen without a progestin) led to what the lay press called an epidemic of uterine cancer. This epidemic was first reported in 1975,[10] and in due time medical experts recognized the cause: unopposed estrogen.

Long before 1975, it was known that menopause could result in the natural occurrence of unopposed estrogen, leading to endometrial cancer. In the 1940s, before the common use of HRT, a pathologist studying the cause of uterine cancer in menopausal women discovered that a change in the endometrium (the lining of the uterus) occurred prior to the development of endometrial cancer. This change was called hyperplasia, referring to an overgrowth of the endometrium. He postulated that this overgrowth was due to estrogen's unopposed stimulation of the uterine lining, and that unchecked hyperplasia would lead to endometrial cancer in about 10 percent of cases.

Before menopause, when a woman is still ovulating and having periods, this buildup of the endometrium occurs every month. Large amounts of estrogen and progesterone are produced by the ovary following ovulation. As these high levels of progesterone and estrogen decline, leading to a period, the endometrium sheds. With the onset of menopause, ovulation ceases and the ovary no longer produces enough progesterone to cause a period. The ovaries make less estrogen, but that estrogen is not offset by sufficient amounts of progesterone. The endometrium may grow too thick—hyperplasia—which is the first stage of abnormal growth that can lead to endometrial (uterine) cancer. Thus, uterine cancer can occur under the natural conditions of menopause, without any supplemental estrogen. Also, estrogen is not a carcinogen (cancer-causing agent), but rather one aspect of a process that under specific conditions can lead to cancer.

After the 1970s' reports of increased endometrial cancer in women on estrogen replacement therapy, estrogen was incorrectly labeled a carcinogen. The use of estrogen plummeted and the FDA recommended that estrogen use be limited to only those women with extreme symptoms, such as continuous hot flashes and night sweats. A form of cancer phobia developed around estrogen, and to this day, some women still associate estrogen with cancer, even after the medical community recognized that an increased risk of uterine cancer is prevented by giving progestin in addition to estrogen. This continuing phobia explains to some extent why 75 percent of menopausal women are not taking hormone treatments, and may, therefore, be putting themselves at greater risk of heart disease, osteoporosis, mood disorders, and cognitive impairments. The old fears about uterine cancer have spilled over to current fears about breast cancer, which I do not mean to dismiss but which must be put into perspective. First, there is no increased risk of uterine cancer as long as progesterone is properly given. The risk of breast cancer from HRT has been overstated. More on this in chapter 9.

Again, menopausal women and their doctors may be making a serious mistake by using progestins on a daily basis to eliminate periods. The body secretes ovarian progesterone in significant amounts for only about ten days during each monthly menstrual cycle. The daily administration of a progestin does not mimic the natural pattern of the body's progesterone secretion and its daily use can cause or exacerbate problems with mood, cognition, and sexuality.

I explain to my patients why I recommend the use of cyclic progestin versus daily progestin, letting them know that they will have a menstrual flow. In many cases, the discussion comes to a screeching halt right there. Extremely intelligent, well-informed women often tell me, "I am menopausal now and I am finished having periods." I certainly understand why they would prefer not to have a menstrual period after they've become menopausal. However, I also know that it is my responsibility to make women aware of the downside of HRT designed to eliminate menstrual periods. In my practice I see many women with depressive symptoms and daily progesterone can cause serious mood disturbances in most of these women. I encourage women who have bad reactions to daily progestin to ask themselves: Would I rather have no periods and be depressed and forgetful? Or would I prefer to have periods but remain emotionally balanced and mentally sharp? Each woman must provide her own answer.

RUTH'S BATTLE WITH PROGESTERONE

Ruth was almost fifty-four when she first started having hot flashes, which were frequently accompanied by palpitations and dizziness. Her gynecologist prescribed daily estrogen plus cyclic medroxyprogesterone added for ten days each month. When she took the medroxyprogesterone, she had a recurrence of hot flashes and became irritable and anxious. In addition, her menstrual periods were irregular. When she described these problems to her doctor, he suggested that she go on a combination of an estrogen and low-dose progestin taken daily. Her hot flashes disappeared, and for two years she had no menstrual flow. "I felt perfectly fine during those two years," she told me. "I loved not having a menstrual period."

However, Ruth began to notice abdominal bloating, pelvic cramping, and breast tenderness. She stopped the estrogen/progestin combination. She felt well for several weeks, though her breast tenderness persisted. Soon she started having headaches and the palpitations returned. Ruth resumed the estrogen/progestin combination; however, this time the estrogen dose was cut in half while the progestin dose remained the same as before. But Ruth's breast tenderness, caused by cystic breasts, became worse, and she decided to stop estrogen to see whether this problem would subside. The tenderness continued on one side, and her hot flashes and insomnia returned, as did symptoms of arthritis that had been better while she was taking estrogen.

Ruth had always been an energetic, active woman, but since going off hormone treatment, she spent her days fighting constant feelings of fatigue. She felt she was in a terrible "Catch-22"—stay on hormones and suffer breast tenderness, bloating, and pelvic cramps; or stay off hormones and suffer from headaches, palpitations, arthritis, and fatigue. Her husband, a physician who'd referred patients to me in the past, suggested that she consult me.

Ruth had extremely low blood estrogen levels, and equally low testosterone. On physical exam, she did indeed have cystic changes in her breasts. Previous experience led me to believe that her daily exposure to estrogen and progestins had produced changes in her breasts that were still causing discomfort. These changes in the right breast were persisting even after the hormones were withdrawn. It was obvious from Ruth's symptoms that she needed to resume estrogen, but I was fearful that estrogen alone would exacerbate the pain in the right breast, since estrogen definitely stimulates growth of breast tissue.

I felt that the best chance we had to solve Ruth's specific set of problems was with proper doses of testosterone along with estrogen. Testosterone has a suppressive effect on the growth of breast tissue. (In the past, testosterone was administered to women who had breast cancer due to its suppressive effects on abnormal growth of breast tissue.) I hoped that low-dose testosterone along with estrogen would allow her to resume estrogen treatment without breast pain. Testosterone has also been shown to increase energy levels in menopausal women. Our goal was for Ruth to resume estrogen without unpleasant side effects.

When I saw Ruth a month after she'd begun the estrogen/testosterone regimen, she was feeling considerably better, her hot flashes had disappeared, and she was sleeping well. The headaches that had plagued her were gone, her arthritis pain had subsided, and her energy had been given a boost. As I had hoped, the tenderness in her right breast was greatly diminished. Testosterone and estrogen were doing their respective jobs.

Now, my major concern was to help Ruth find a progesterone regimen she could tolerate. My experience told me that micronized progesterone given cyclically for ten days each month might have the best chance of success. Still, I was not overly optimistic, since the anxiety and irritability she had previously experienced on synthetic progestins made it clear that she was sensitive to the negative effects of progesterone. My fear was confirmed by a phone call from Ruth telling me that she'd stopped micronized progesterone because, in her words, "It made me sick. I felt miserable."

When I saw Ruth several weeks later, we discussed strategies to deal with her progestin sensitivity. Daily and cyclic progestins had not worked for her in the past, so I suggested vaginal micronized progesterone for ten days each month. But Ruth did not feel good about this option. I offered another strategy—to reduce the frequency of cyclic progestin from monthly to every three months. I did not believe this would reduce her progestin sensitivity or eliminate her PMS-like symptoms, but she would only experience these side effects every third month. Of course, there was a caveat. Under conditions of reduced progesterone exposure, it was crucial that we regularly monitor the thickness of Ruth's uterine lining. Ultrasound can measure endometrial thickness in a relatively painless procedure performed in a gynecologist's office or in the diagnostic radiology department of a hospital or clinic. However, ultrasound is not as precise a diagnostic method as endometrial biopsies, which can entail some discomfort. Unfortunately, there is no perfectly satisfactory and painless method of monitoring the endometrium.

One sensible approach in Ruth's case, and for other women taking progesterone every three months, is to have ultrasounds every six months and whenever abnormal vaginal bleeding occurs. Endometrial biopsies, the only surefire way to diagnose hyperplasia, would be performed whenever the physician or radiologist noticed an increase in endometrial thickness or any other abnormality found on ultrasound examination. This would allow for early diagnosis and curative treatment.

Ruth chose to take medroxyprogesterone every three months, and indeed, it produced far fewer side effects than micronized progesterone. She had escaped her "Catch-22," while doing everything possible to protect the health of her uterus. Ruth's story illustrates several important points. A key one: The daily use of a combination of estrogen/progestin is very popular with physicians and menopausal women, since it usually eliminates menstrual flow. However, this regimen can come with a very high price tag in the form of serious physical and/or emotional side effects. The only solution is custom-tailored, rationally designed hormone modulation.

DEBORAH'S STORY: HORMONE MODULATION IN ACTION

Just after she turned forty, Deborah was diagnosed with a large uterine fibroid, the cause of her excessive menstrual bleeding. She had surgery to remove her uterus, but her ovaries were left intact. Seven years after this surgery, Deborah consulted me because she was experiencing pounding headaches and bouts of depression, during which she'd spend whole days on her couch. Her moods could swing wildly, and at times, she'd experience an excess of energy, taking on too many projects until she would find herself going full tilt at 3 A.M. Along with these mood swings, her weight would fluctuate by thirty pounds. Though she was not particularly obese, weight loss was a constant concern. When she was able to exercise and lose weight, her mood improved, but when she was down, she had no energy to exercise, sending her into a downward emotional spiral. Deborah had a family and personal history of depressive episodes, so she was one of those women who are vulnerable during "the change." Antidepressant medications were not working.

Blood tests revealed that Deborah had slightly elevated FSH, consistent with perimenopause. Her estrogen levels fluctuated between normal and low, while her total and free testosterone levels were consistently low. Since hysterectomy can sometimes disrupt the blood flow to ovaries left intact and bring about early menopausal changes, I suspected this was happening to Deborah. I started her on estrogen, and after several weeks, we added low-dose methyltestosterone. When I saw her a month later, her moods had become far more balanced. She no longer experienced those extreme hyperactive highs or depressive lows. The fatigue that had plagued Deborah for much of her life, in fact, was finally gone, and she was having fewer headaches.

However, after several months on estrogen Deborah began to complain of breast tenderness, so together we decided to reduce her estrogen dosage. Unfortunately, on the reduced dose, her depression and fatigue returned. One of the problems Deborah had noticed on the higher dose of estrogen was fluid retention. We decided to resume her higher dose of estrogen, but this time we added a low dose of a diuretic drug. When I saw her again a few weeks later, she was smiling. As she entered my office, her demeanor was cheerful and lively. "There is no question," she told me, "that my mood and energy are much better on the higher dose of estrogen, and on the diuretic, breast tenderness is not a problem." I was astounded by the extent of the change in Deborah, who'd been overweight and felt sluggish for years.

"I have lost ten pounds," she told me. "I'm eating healthier, my diet consists of fruits, vegetables, fish, and pasta. Now that I'm not depressed, I can stick to a diet. I go to the gym three times a week and when I walk out, I feel so much better. I love my husband, I love my children, I love my life."

Deborah's story is not atypical. I see many perimenopausal women with both a family and personal history of depression who in the past had required antidepressants. As hormone levels shift, these symptoms either reappear or worsen. In Deborah's case, when her declining estrogen and testosterone levels were corrected using hormone modulation, her symptoms literally disappeared, and she no longer needed antidepressants.

Deborah's case however differs from those of women who have their ovaries removed at the time of hysterectomy. After knowing Deborah for two years, she revealed how close she had come to losing her ovaries. Her surgeon specialized in ovarian cancer, and he frequently recommended that patients undergoing hysterectomy have their ovaries removed as a preventive—even if they did not have existing cancer. Deborah agreed, but on the day of surgery, while she lay on the operating table prior to being anesthetized, she had second thoughts. Tears filled her eyes and she began to weep. Much to his credit, the surgeon recognized Deborah's sadness and asked her whether she was sure about her decision. "If you don't want to do this, just say so," he told her. "It doesn't have to be done. It's totally your decision."

Deborah opted to have her uterus but not her ovaries removed. I am certain that Deborah's mood difficulties were much easier to treat because she had intact ovaries. While her ovarian function declined during perimenopause, and could not totally protect her mood, the small amounts of "endogenous" ovarian hormones—those made by her own body—conferred some degree of protection. This made it easier to treat her with supplemental hormone medications. I have treated a number of women similar to Deborah who had their ovaries removed during hysterectomy, and their cases were much more complicated and difficult to treat.

Doctors who say, "You don't need your ovaries" based on the rationale that you are near or past menopause may be wrong. The minuscule amounts of ovarian hormones produced are better than none at all. I also subscribe to the belief that organs of the body are not expendable unless they are diseased or destructive of our health and well-being. One such exception: Women with the gene for breast and ovarian cancer who are at high risk for life-threatening disease may want to consider prophylactic surgery.

HORMONAL PREVENTION OF HEART DISEASE

Of all the positive health effects of HRT, unquestionably one of the most important is the effect of estrogen in preventing heart disease. There are numerous reports in the scientific literature that reveal that long-term estrogen deficiency leads to an increased risk of heart disease. And we must bear in mind that heart disease is the leading cause of death in menopausal women. At least 350,000 women over forty-five die annually from heart attacks, representing 48 percent of all deaths for women in that age bracket. This represents more than eight times the annual number of deaths caused by breast cancer in all ages.

In an era of fear about breast cancer, it is vital that women confront one simple fact: For the vast majority, heart disease is a far, far greater threat.

There is no question that estrogen protects the heart. As women enter menopause, their cholesterol levels rise, their "good" (HDL) cholesterol declines while their "bad" (LDL) cholesterol climbs, their arteries become more vulnerable to the buildup of plaque, and they become more prone to the abdominal fat deposits that are linked to cardiovascular risk. A primary reason for all these unhealthful changes is the precipitous decline of estrogen. Which is not to say that diet, exercise, and stress don't play a role in heart health, just that estrogen loss is an undeniable factor in women's increased risk of heart disease as they age.

Estrogen replacement therapy reduces heart attacks in menopausal women up to 50 percent. I will offer more detail on this finding in chapter 9. But it is based on many so-called "cross-sectional" studies that evaluate a population of women who are either taking or not taking hormones at a fixed point in time. Such studies provide strong but not foolproof evidence, since other factors related to hormone use—like income or education—may be the real underlying reason for reduced heart risk. Although many such studies tried to account for these factors, they still may not be definitive.

The Women's Health Initiative Study

One way to overcome this problem is to conduct a "prospective" study, one that follows the same population over a long period of time, with one portion of the women receiving hormone treatment while another receives a placebo. Such a study—the "gold standard" of clinical research—can better control for other factors, like economic status, and ultimately determine with greater certainty and accuracy whether hormone treatment reduces the risk of heart disease.

The ongoing Women's Health Initiative Study (WHIS) is such a study, and it will take roughly a decade to complete. The goals of this study are to examine the effect of standard hormone-replacement therapy (HRT) on the incidence of heart disease, osteoporosis (thinning of the bones), and Alzheimer's disease in a broad sample of women. Admirably, a distinct effort has been made to include minority groups and women of varying social and economic backgrounds. However, I am concerned about one aspect of this study. I fear that it will not answer many pressing questions regarding the use of progestins in women receiving estrogen therapy. Women in the WHIS will get either a daily combination of estrogen plus synthetic progestin (medroxyprogesterone) or they will get a placebo. (Women who don't have a uterus will get either estrogen alone or a placebo.)

There are at least two serious omissions in the WHIS. The study will not evaluate women who get cyclic progestin—for only ten days each month—and it will use only one type of synthetic progestin. In my view—and this is supported by many studies—progestins can offset the benefits of estrogen to both the heart and the brain. They are also responsible for most of the negative side effects associated with standard HRT. Different types of progesterone (such as natural micronized progesterone) and regimens that include progestins for only ten days each month protect the uterus and may do so without sacrificing the benefits of estrogen for mental and cardiovascular health. This is the type of approach I have taken with my patients.

HERS: The Truth About Hormones and Heart Disease?

Recent findings lend credence to my concerns about the need to study new options in hormone modulation. A 1998 study published in the *Journal of the American Medical Association* tracked 2,763 menopausal women at twenty medical centers, all of whom had a history of heart disease.[11] This is known as the HERS study. The women, who were followed for four years, were either treated with a daily combination of estrogen plus synthetic progestins, or with a placebo. During the four-year follow-up, treatment with estrogen plus a progestin did not decrease the incidence of heart attacks or the number of mortalities in postmenopausal women with previous heart disease. In year one there was a 52 percent increase in the incidence of heart events. While the overall risk for heart attacks was not reduced there was a significant trend over the four-year period. This trend revealed fewer heart attacks occurring in the HRT group compared to the placebo only in the third and fourth years.

The HERS study shook the strong belief of most physicians that estrogen therapy can prevent heart disease in menopausal women. But here again, the study used an HRT regimen with daily doses of synthetic progestin—the type most likely to offset the cardiovascular benefits of estrogen. Furthermore, since all the women in the study had a history of coronary artery disease, it is impossible to say if the results apply to healthy women who want to prevent heart disease.

It is known that estrogen/progestin treatment may induce abnormal clotting, one of the first serious side effects identified with the use of "the pill," which contains such a combination. When this combination is given to a woman with existing cholesterol plaques that clog her coronary arteries, it can lead to the formation of clots in those arteries—a potential cause of heart attacks. The arteries of the heart are not the only place where an estrogen/progestin combination may cause clotting problems. In fact, the women in the HERS had a threefold increase in "thromboembolic events." A thromboembolism involves clot formation in a vein. The clot may break off and travel through the circulation to the lungs, or to the brain, causing a stroke.

In the wake of the HERS report, I have not been the only expert to suspect that the failure to find a heart-protective effect of HRT is due to the continuous daily administration of synthetic progestins with estrogen. These suspicions are supported by studies in monkeys, showing that estrogen alone decreased cholesterol plaques in coronary arteries by 72 percent. However, the plaque formation in the coronary arteries of monkeys treated with a combination of estrogen plus synthetic progestin (medroxyprogesterone) did not differ from untreated controls.[12]

In an earlier study in monkeys the administration of both estrogen alone or estrogen plus natural progesterone given cyclically produced a

50 percent reduction in coronary artery plaque formation.[13] The message from these studies would seem to indicate that natural progesterone does not counter the positive effects of estrogen, while synthetic medroxyprogesterone does.

Further animal studies showed that estrogen alone boosted blood flow to the heart by about fifty percent, whereas the combination of estrogen with any one of three synthetic progestins abolished this increase. These important animal studies support wholly the view that continuous progestins can wipe out the heart-protective benefits of estrogen.

Estrogen Prevents Coronary Artery Disease

There is another reason that giving estrogen to women with heart disease does not significantly reduce the number of heart attacks. From both animal and human studies it appears that high estrogen levels prevent the development of coronary artery disease (CAD) in which cholesterol plaques block blood flow. If the plaques are already there, estrogen treatment does not appear to reverse the process. Therefore, it is important to make sure that sufficient levels are maintained to prevent CAD before it develops.

Estrogen: A Cure or a Prevention?

In August 2000, a group headed by Dr. David Herrington of Wake Forest University reported on a study of 309 older women with known heart disease who had been menopausal for a protracted period.[14] These women were treated with either estrogen alone, a combination of estrogen plus progestin, or a placebo. Using sophisticated computer techniques the women were followed for three years to detect subtle changes in the arteries of the heart. Overall Dr. Herrington found that treatment with estrogen by itself or estrogen plus a progestin did not slow the progression of heart disease in these women who already had established narrowings in the arteries to their heart. These studies have been interpreted by some physicians to mean that we were mistaken when we believed that estrogen prevented heart disease. What these studies show is that estrogen alone or estrogen/progestin treatment does not reverse coronary artery disease. These results could be interpreted to mean that estrogen treatment does not prevent heart disease, but I believe such an interpretation would be a mistake. What these studies do not answer is whether estrogen if administered early on in menopause can prevent the development of heart disease. All of the "cross-sectional" case report studies support the hypothesis that estrogen prevents the development of plaque formation in the arteries of the heart. Also supporting this hypothesis is the observation that premenopausal women with high estrogen levels die infrequently (13 percent) from heart disease. After menopause when estrogen levels plunge, this changes drastically, and deaths from heart disease increase to approximately 50 percent.

The Search for Heart-Healthy Progesterone

I've argued that some form of progesterone is necessary to protect the uterus from abnormal growth. But is there a heart-healthy progesterone? Fortunately, animal research suggests that natural progesterone won't wipe out the cardioprotective effect of estrogen. At the Oregon Regional Primate Center, researchers showed that estrogen plus natural progesterone given to estrogen-deficient monkeys protected them against spasms of the coronary arteries. Such spasms may lead to heart attacks. Estrogen alone also protected against these spasms, but the combination of estrogen and synthetic progestins failed to do so.

The Postmenopausal Estrogen/Progestin Interventions (PEPI) Trial studied 875 postmenopausal women, examining the effect of various hormone combinations on the so-called good cholesterol.[15] Daily estrogen plus cyclic micronized progesterone produced greater increases in HDL levels than did daily estrogen plus either cyclic or daily synthetic progestins. These higher HDL levels have been shown in a plethora of studies to protect against heart disease.

In summary, there's no doubt that estrogen can protect the heart, but HERS suggests that the addition of daily progestins significantly reduces this benefit. Now the experts must determine what form of progesterone will prevent uterine cancer without blocking the heart-saving effects of estrogen. Based on all the available evidence, I believe that the best choices are micronized progesterone administered as a pill or vaginal suppository, preferably on a cyclic rather than daily basis. If synthetic progestins are given, cyclic administration might be better for the heart *and* the brain. That said, we need those large-scale studies to nail down which approach is ideal. It is terribly unfortunate that the Women's Health Initiative Study (WHIS) trial, which will follow 27,000 women, will not test any of these regimens—only daily estrogen/progestin. I hope that a valid prospective study will soon be undertaken to answer these questions in the foreseeable future.

The HERS finding has spurred a new round of questioning hormone therapy. At worst, this flawed study suggests that HRT will not help a woman who already has heart disease. I am frankly concerned that women will again be dissuaded from taking estrogen, and countless will die prematurely of heart disease, osteoporosis, or Alzheimer's disease, and live with a greatly diminished quality of life. Until we have optimal clinical trials—ones that investigate a range of options—women must still decide whether to pursue hormone modulation. They must not do so in a media-inspired atmosphere of fear and confusion. Twenty-five years ago, the use of estrogen plummeted when it was blamed for uterine cancer. The paranoia that followed is still plaguing us today, and millions of women who might benefit from HRT are dissuaded. We ought to learn from past mistakes, not repeat them.

In April 2000 my fears regarding the WHIS were further confirmed when some initial early results were leaked to the press. While the data were not available for examination, the press release stated that the study was finding that estrogen did not protect against heart disease. This is a very damaging conclusion. All women with a uterus in the WHIS were treated with Prempro, a combination of estrogen and a progestin, medroxyprogesterone. This combination is not estrogen. The only women in the study who received only estrogen were women who had a hysterectomy, thus significant portions of the data were contaminated by the administration of a progestin, which may have acted in a major way to inactivate the positive effects of estrogen on heart disease. Only when the study is completed and analyzed in at least 2005 will we really have a full understanding of the effect of estrogen alone on heart disease prevention. I would normally refrain from commenting on such incomplete data but since it is receiving wide coverage in the media I feel it is necessary to warn women against accepting invalid data reporting.

In August 2000, the *New England Journal of Medicine* published a study entitled "Trends in the Incidence of Coronary Heart Disease and Changes in Lifestyle and Diet in Women."[16] In the Nurses Health Study, 85,941 women ages 34–59, who had no previous diagnosed cardiovascular disease, were followed from 1980 to 1994. During that period the incidence of coronary disease declined by 31 percent. This decline was the result of a variety of factors. A multivariate analysis suggests that changes in cigarette smoking, diet, and postmenopausal hormone use statistically explain the reduction in the incidence of coronary disease. There was a 175 percent increase in postmenopausal hormone use. The fact that this was a prospective study with a large sample and a high rate of follow-up with detailed information collected on diet and lifestyle factors make the results of this study extremely powerful.

DIET, EXERCISE, AND MENOPAUSE

While hormone modulation is critical for the maintenance of good health, it cannot do the job alone. Diet and exercise are essential requirements for the prevention of such diseases as osteoporosis, coronary artery disease, and depression. I embrace a whole-person, whole-body approach to treating problems associated with menopause, and as such, I encourage you to pursue a healthful, moderately low-fat diet and to get daily physical exercise.

Earlier in this chapter I told the story of Deborah, who became perimenopausal at an early age and was stricken with depression and flagging energy. Deborah was simply too depressed to get much physical activity—she could barely get off the couch to make a cup of coffee. Her inactivity contributed to her weight gain, and she no longer seemed able to break the worsening cycle of emotional misery and physical lethargy. Hormone modulation had a powerful antidepressant effect, and it finally sparked the psychophysical changes that enabled her to take up an exercise program, then to change her diet and lose weight. A positive cycle was set in motion, and the result for Deborah was a newfound energy and a joyous quality to her days. Exercise itself is a mood enhancer, and a balanced, nutritious diet also supports psychological well-being.

During the menopausal passage, many women get stuck in such vicious cycles, and the best way to work out of the thicket of ill-health is to follow several paths at once: a sound diet, an enjoyable exercise program, and hormone modulation when indicated by symptoms and blood tests.

Depression and a lack of exercise have been found to be major predictors of weight-loss failure. A recent study reported that there was a strong positive correlation between obesity and depression. Once Deborah's depression lifted, she was able to exercise and lose weight. My clinical experience confirms that it is almost impossible for depressed people to successfully follow an exercise and weight-loss program. The precise relationship between depression and exercise is not well understood, but recent studies reveal that exercise influences mood by boosting or balancing neurotransmitters in the brain.[17] Exercise clearly boosts one set of neurohormones, the mood-enhancing and pain-killing endorphins.

Unfortunately, only one half of middle-aged women exercise regularly, and less than one quarter follow the NIH guidelines, which suggest moderate exercise for an hour a day, six days per week.[18] Sound research and my own experience with patients confirms that diet and exercise help keep women's moods balanced during and after the menopausal passage.

During the menopausal transition, changes occur in a woman's body composition. She undergoes a decline in bone mineral content, a loss of lean body mass, and an increase in total fat, particularly abdominal fat. The rapid decline in estrogen plays a role in these changes in body composition, and diet and exercise can offset them to some extent. A particularly significant change is the decrease in lean muscle and fat-free body mass in the menopausal woman. The muscles she uses during exercise expend a large number of calories. Muscle even uses calories at rest. When lean muscle is lost, because of declining testosterone levels, it often is replaced by fat. An equivalent amount of muscle burns seven times the number of calories as fat. So when this muscle mass decreases, energy expenditure goes down and weight goes up—especially if the woman does not cut back on calories. The answer? It's a combination of a calorie-reducing diet along with appropriate estrogen and testosterone treatment. Testosterone may partially offset the loss of muscle mass and rise in total body fat.

Both exercise and hormone modulation are critical for the prevention of osteoporosis (thinning of the bones), which can vastly increase women's risk of severe bone fractures. Studies at Washington University School of Medicine in St. Louis have shown that exercise plus HRT are more effective than either exercise or HRT alone in increasing bone mineral density, protecting against the development of osteoporosis. I will cover the prevention and treatment of osteoporosis in chapter 9.

Women need to exercise and maintain a reasonable body weight in order to maintain normal cholesterol levels. A recent study, reported in the *New England Journal of Medicine,* evaluated the effect of exercise and diet on HDL (good cholesterol) and LDL (bad cholesterol) levels in menopausal women. Only the combination of dietary modification plus exercise resulted in significant reductions in LDL—the "bad" cholesterol.[19]

We also know that estrogen-based hormone treatments can reduce LDL and raise HDL. While HRT can be critical for maintaining cardiovascular and overall health, it can't do the job alone. Clearly, women need a sound diet and exercise program to prevent coronary artery disease, osteoporosis, and depression. More recent studies suggest that we must add Alzheimer's disease, certain cognitive impairments, sexual dysfunction, Parkinson's disease, some autoimmune diseases, and even colon cancer to this growing list.

Several years ago, I realized that I needed to lose weight and exercise more, so I joined a local gym. One day at the gym, I was pleasantly surprised to run into Sally, a friend I hadn't seen for several years. Sally was menopausal and on HRT, and she seemed to have been emotionally and physically transformed. Like me, she had recognized just how important it was that she exercise to maintain a normal weight and build strong bones and muscles. As a result of her health program and vigorous exercise, Sally looked fifteen years younger than her age, and she was optimistic, vibrant, and full of joy.

ALTERNATIVE MEDICINE AND HORMONE MODULATION

Since only 25 to 30 percent of menopausal women use HRT, the majority are not being treated. There are many reasons for these statistics, including fear of breast and uterine cancer, and concerns about the side effects of standard doses of HRT, such as weight gain, fluid retention, and mood changes. Some of these concerns are overstated, yet many women are searching for alternative treatments to HRT. These include nutritional supplements, herbal and homeopathic remedies, and physical approaches such as diet and exercise. In addition, women are turning to some psychopharmacologic medications, including antidepressants and antianxiety drugs.

Soy Foods and Products. One of the most popular alternatives to HRT has been the use of soy, whether from food sources (tofu, tempeh, and miso) or supplements. Soy is one of several plants containing compounds called phytoestrogens, which have estrogen-like actions. Soy contains isoflavones such as daidzein, an established phytoestrogen. An Australian study examined the effect of a daily supplement of soy flour on hot flashes in fifty-eight postmenopausal women. After about six weeks, hot flashes decreased by about 40 percent.[20] While this is an important finding, women with hot flashes must compare the 40 percent reduction with the near 100 percent reduction that occurs with

HRT. Decreases in blood cholesterol levels and urinary calcium excretion seen with estrogen did not occur in the soy-treated women. Such blood changes protect against heart disease and osteoporosis.

In another study carried out at the National Institutes of Health Sciences in North Carolina, about 100 postmenopausal women were evaluated to determine whether a soy-supplemented diet would produce estrogenic responses. Half the women were given supplemental soy, while the other half were instructed to follow their usual diet. On the basis of published estimates of the estrogenic potency of soy isoflavones, the soy program was expected to have estrogen-like effects, but they were not seen.[21] HRT suppresses the pituitary gonadotrophins, FSH and LH, and it increases sex hormone binding globulin, but these changes did not occur with the soy supplement. At most, the women on soy had a minor estrogenic effect on vaginal cells, though it was not statistically meaningful.

Advocates of soy as an alternative to HRT suggest that its constituents are so weakly estrogenic that they have no negative effects on breast or uterine tissues. They speculate that soy may even reduce breast cancer by blocking estrogen receptors, though the evidence for that claim is, to date, insubstantial. On the other hand, the very weakness of the estrogen-like compounds in soy explains why it may be only somewhat effective in the treatment of severe menopausal symptoms. Women can certainly try soy products, which are otherwise healthful, and determine for themselves whether their symptoms—including their emotional or cognitive symptoms—are relieved. To date, there are no data proving that soy-based phytoestrogens can alleviate the mood, cognitive, or sexual symptoms of menopause.

Women with mild symptoms who are helped by soy or other alternative options may feel they do not need hormone modulation. However, I encourage every woman to bear in mind the other long-term benefits of hormone therapy—reduced risks of heart disease, osteoporosis, Alzheimer's, and other diseases. Depending on your unique medical history and sense of your own biological individuality, you may wish to commence with hormone modulation even in the absence of severe symptoms, or with symptoms that are relieved by alternative medicine.

For instance, two women with no appreciable symptoms at menopause may make completely different decisions about hormones. If one has a family history of heart disease and osteoporosis, she may wish to initiate hormone therapy, while the other woman, with no such family history, may decide not to take hormones.

St. John's Wort for Mood. St. John's Wort (*Hypericum perforatum*), a flowering perennial plant, has been used medicinally for thousands of years. Recently, there has been an explosion of publicity about St. John's Wort as a treatment for mild to moderate depression. Pharmacologic studies indicate that the antidepressant action of St. John's Wort is explained by its influence on neurotransmitters, serotonin, norepinephrine, and dopamine. This herbal remedy has been used extensively in Europe, particularly in Germany. In European clinical trials, St. John's Wort compares favorably to tricyclic antidepressants, such as amitriptyline or imipramine. (And the side-effect profile of the herb was clearly superior to that of the tricyclic drugs.)[22] A statistical overview of more than twenty clinical trials, published in the *British Medical Journal*, confirmed that St. John's Wort can be effective for mild-to-moderate depression. The most common side effects of St. John's Wort were similar to that of a placebo and included a small percentage of patients with gastrointestinal symptoms, dizziness, confusion, tiredness, and sedation. There have been rare reports of photosensitivity.

Now under way, a multicenter clinical trial supported by the National Institute of Mental Health is exploring the efficacy of St. John's Wort for mild depression. A comparison will be made between Zoloft (sertraline), St. John's Wort, and a placebo. Data from this study will be available sometime in the year 2000. It will be the first major test of St. John's Wort in the United States. Only if test results are positive for St. John's Wort will most physicians be willing to recommend this herbal remedy for depression. However, patients will continue to self-medicate with herbs for the relief of symptoms, assuming that these products are safe because they are "natural." The European trials indicate that the herb itself, taken in proper dosages, is indeed safe. But I've spoken with psychiatrists who've had patients develop serious overdose complications when the herbal medicine was used in combination with SSRI antidepressants, such as Prozac and Zoloft. Responsible practitioners of alternative medicine know that these interactions are unsafe, and they inform patients never to combine St. John's Wort with conventional antidepressants. But I'm concerned about people who grab the herb off the health food–store shelf without proper guidance. At the moment, the United States has no regulatory system for herbal products. Physicians and patients alike must educate themselves about the efficacy and contraindications of herbal remedies.

Ginkgo for Memory and Cognition. Another popular plant extract is Ginkgo Biloba, which has been approved by the German health authorities for the treatment of Alzheimer's disease and other cognitive impairments. However, there are twenty-four different brands of Ginkgo Biloba available in the United States, and this presents doctors and patients with a quandary. What brand is best? Quality control is often lacking in U.S. herbal products, since they are no longer under FDA supervision. This is especially unfortunate, since Ginkgo does appear to have benefits for people with memory loss and/or dementia. An analysis of more than fifty articles examining the effectiveness of Ginkgo Biloba on the cognitive function of patients with Alzheimer's disease revealed only four studies that met the requirements for a clear diagnosis of Alzheimer's. However, in those four studies, which compared Ginkgo to a placebo in over 400 subjects, researchers found that the herb produced a small but significant improvement in cognitive functioning. Ginkgo was associated with no significant side effects, other than two cases of bleeding complications.

Scores of European studies have confirmed the value and safety of Ginkgo, but some physicians remain reluctant to widely recommend its use, and that of other similar alternatives, until American investigators have conducted clinical trials, standardized dosages, and ensured quality control. I don't dissuade patients from pursuing alternative approaches, as long as they consider every angle, inform themselves, and find out about adverse effects or drug interactions.

COMPLEMENTARY MEDICINE: BEST OF BOTH WORLDS

I encourage patients not to develop an "either/or" approach to alternative and mainstream medicine. Often, patients can combine the best of conventional medicines with alternative approaches, in what is commonly known as "complementary medicine."

For instance, menopausal women with cognitive symptoms may wish to try Ginkgo to revive their mental sharpness. That does not mean they should automatically reject hormone treatments that may relieve their hot flashes, ease their depression, prevent heart disease, and keep their bones strong and healthy. Also, St. John's Wort relieves depression through mechanisms that are not entirely dissimilar from SSRIs—it appears to bolster serotonin functions in the brain. Women with declining estrogen may lose serotonin receptors that enable them to respond positively to any antidepressant—including St. John's Wort—that bolsters serotonin. While speculative, some menopausal women may do better with St. John's Wort when they are also taking estrogen to optimize their serotonin receptors.

Finally, there is no doubt that general health measures, including a sound low-fat diet, careful vitamin and mineral supplementation, physical exercise, stress management, relaxation techniques, and pursuing a meaningful spiritual life are all contributors to women's health. None of these approaches contradicts hormone modulation. Indeed, they complement rational hormone therapy designed to rebalance a woman's biochemistry for the optimal health of body and mind.

HORMONES AS "NATURAL" MEDICINE

Many women feel they can go through menopause without using hormones, and certainly there are some who can. But I repeatedly hear the statement, "My mother and grandmother didn't take hormones, and they did just fine." But this is an overgeneralization, since many mothers and grandmothers did not do wonderfully at menopause; they just didn't share their miseries.

One reason that menopause was of less concern to our ancestors relates to changes in life expectancy. When women did not make it to their sixties, living beyond menopause was less of an issue. Perhaps the body was not naturally designed to outlive for decades the decline of ovarian function. Of course, with modern medical advances, women entering the new millennium can be expected to live well beyond eighty. Consequently, without hormone treatment, their bodies will endure more than thirty years of estrogen and testosterone deficiency. Is this really what nature intended?

As the baby boom generation ages, 21 million women will enter menopause. In Dr. Susan Rako's landmark book *The Hormone of Desire,* she writes, "The natural way for me was to develop testosterone deficiency at age forty-eight, the symptoms of which, yielded not at all to the ministrations of the best homeopaths, Chinese herbalists, acupuncturists, and healers of eclectic persuasion . . ."[23] Dr. Rako was only able to resolve the symptoms of testosterone deficiency by replacing the missing hormone. Her experience reinforced my own observation that when estrogen, testosterone, and progesterone become deficient after menopause, the most natural way to treat women's symptoms is to replace those hormones. The body will respond similarly once a deficient hormone is replaced, as long as the replacement mimics the natural quantity and quality of hormones as closely as possible.

Let me explain why hormone substitutes, whether in the form of mainstream drugs or nutritional supplements, may not always be preferable to hormones themselves. A woman's bones, heart, sex organs, and brain all contain hormone receptors that attract circulating hormones to the specific site in the body where they are meant to do their job. These receptors—proteins that sit on or near the cell surface—act like magnets, pulling the hormone molecule into the cell. Such receptors are present on cells in areas of the brain that require regulation of particular functions by particular hormones. For example, estrogen receptors are present on brain structures that control sleep, mood, memory, learning, and sexual function. Testosterone receptors are present in areas controlling sex drive, sex function, mood, and energy. These receptors are like air-traffic controllers, safely directing hormones to the specific landing sites in the brain and body where they belong.

When certain drugs are taken—whether mainstream (i.e., tamoxifen, raloxifene) or alternative (i.e., soy phytoestrogens)—hormone receptors may not respond in the same fashion that they do to natural endogenous hormones. These agents may bind to receptors in the brain or other organs and exert an effect, but in many instances, this binding is weaker, less potent, or less selective. The agent may act in parts of the body where the hormone would not be bound, and the result may be adverse side effects. Obviously, I do not mean to suggest that synthetic drugs are never useful or effective. Rather, when natural substances such as testosterone or estrogen can achieve optimal effects, they are often preferable.

Hormone modulation strives to deliver hormones in as natural a manner as possible. As I have emphasized, that means returning blood hormones to normal premenopausal levels, not to super-high levels. Often, that also means using hormone preparations closer to their own hormones (i.e., natural progesterone) and schedules that more closely recreate a woman's natural hormonal flux (i.e., cyclic rather than daily progesterone). Certain negative effects that occur with standard HRT may result from the fact that it does not sufficiently mirror a woman's natural hormonal profile.

When people comment that giving hormones after menopause is "unnatural," I understand their assumption—that if "nature" meant women to stop making hormones at a certain age, we should leave well enough alone. But as a physician, I must be concerned about my patients' health and I cannot accept that "nature" meant so many women to experience terrible emotional, mental, and sexual distress that ruins their quality of life. I can only assume that "nature" could not have predicted that our medicine would advance to such heights that women would routinely live into their seventh and eighth decades. Now that we've learned how to keep people alive with sophisticated medical technology, it's our responsibility as doctors to keep them psychologically balanced, productive, and healthy, as well.

THE NEW MENOPAUSAL WOMAN

I feel extremely fortunate to be practicing medicine at a time when women are taking a much more active role in their own medical care. Today's woman—generally more confident, assertive, and well informed—arrives at the doctor's office prepared for a two-way conversation about her care. This is particularly true of the woman at menopause who must decide about hormone replacement.

When it comes to decisions about menopause, the doctor-patient relationship is perhaps more pivotal than ever. The reason is tinged with irony: There is more information out there than ever before. Media coverage of issues about menopause and hormones is so widespread that one can hardly pick up a magazine or turn on the TV without being told about the latest treatments to relieve menopausal complaints. Books about menopause-related issues seem to fly off the bookstore shelves. But so much information is contradictory that the average woman cannot possibly make clear-cut decisions. She would have to delve into the medical literature to try to untangle the confusing thicket of conflicting information. While I encourage women to do just that—to conduct their own research and become informed—I also believe that doctors can and should be professional guides who help them make sound decisions on their own behalf.

Unlike her predecessors, the "new" menopausal woman is not willing to simply be given a prescription and be told to come back in six months. She does educate herself, arriving at her doctor's office armed with questions, looking for answers. If her hormonal therapy produces side effects, she expects her physician to listen to her complaints and respond intelligently. No longer is she willing to accept the answer, "You'll just have to put up with it," or even worse, "It's all in your head."

The latter comment is a kind of insult. The real meaning is, "You're just imagining these problems, making matters worse with your own negative thoughts." I can tell you just how rare it is for any woman to be suffering imagined or exaggerated symptoms of menopause. But you can redefine the phrase in a way that makes it true. Namely, peri- and postmenopausal women suffering from blue moods, outright depression, memory problems, fuzzy thinking, and sexual dysfunction may indeed have a problem. But it is not "all in their heads" as some women have been told. It's no fantasy and it's no reason for shame. I tell my patients, "You have nerve cells, neurotransmitters, and blood vessels in your head that are being affected by your hormones, or lack of them." Doctors must understand the impact of hormones on the brain in order to correctly interpret their patients' very real symptoms. And menopausal women must themselves be aware of this hormonal impact, knowing that it is common, understandable, and no sign of personal weakness.

When physicians listen to their patients' concerns with compassion and understanding, women can honestly confront their menopausal symptoms, including those that affect the mind, without guilt, shame, or fear. Then, doctors must know how to modify hormone regimens so that they are most effective, safe, and cause the fewest possible side effects.

Many of the adverse effects associated with standard HRT may be avoided if hormone regimens are flexibly designed and are administered so that they restore proper hormonal balance in the brain and the other organs and tissues of the body. That's what hormone treatment should be all about.

Hormone regimens for menopause are not magic. They require a partnership between doctors and patients, in which both pay attention to all of the patient's symptoms and concerns. The physician may present an option that can most effectively treat all the patient's symptoms. Then he or she raises or lowers dosages, and if necessary, suggests alternate routes of administration, seeking to provide maximum health benefits with the fewest side effects. All of the woman's health goals must be considered—prevention of heart disease, cancer, osteoporosis, and Alzheimer's, as well as relief of menopausal symptoms, including those that affect the mind. Clearly, there is no easy blueprint for this approach. Each woman must receive highly individualized and respectful treatment—what I call hormone modulation.

8

The Art and Science of Hormone Modulation

DIAGNOSIS AND TREATMENT

When is hormone modulation appropriate? The answer is whenever there are indications that a shifting hormonal status has triggered a set of symptoms, particularly mood, memory, cognitive, and sexual disorders. In previous chapters, you learned that women with such symptoms, whether associated with PMS, postpartum changes, perimenopause, or menopause itself, can be treated with a custom-tailored hormone therapy that is often more effective than treatments tried in the past. Effective therapy may be possible with hormone modulation alone, or hormone modulation with antidepressants, antianxiety agents, or other conventional medicines. One cue to proceed with hormone testing occurs when mood or memory disorders do not respond to standard treatments, such as antidepressants like Prozac.

I've offered considerable information about the when and why of hormone modulation. In this chapter, I offer some sense of the how of hormone modulation. How is it done? How do clinician and patient make rational decisions when so many options are available? How can you fashion a regimen of hormone therapy that is individually tailored, safe, and effective? How do you start and how do you monitor your progress in ways that make sense?

The best way I can answer these questions is to offer you case histories—much the way physicians offer cases at medical conferences to illustrate how they handle and overcome clinical problems. Consider this chapter on the art and science of hormone modulation as a series of case presentations to help you better understand how you and your doctor can properly treat your hormone-related symptoms of mood, memory, cognition, and sexuality. I will share with you some of the techniques of diagnosis I use to assess women's hormonal problems, and guide you through my thinking as I help patients develop successful programs of hormone modulation.

Bear in mind that you will have to work closely with health care professionals to develop a customized program of hormone therapy. This requires cooperation and communication with not one but usually several physicians, and it's up to you to identify a medical facilitator and at least one specialist who will work with you. I fully explain how you can embark on this process in chapter 10, "Taking Charge of Your Hormones."

A WOMAN'S HORMONAL HISTORY

I begin my search for a diagnosis by taking a detailed medical history: Did the patient develop according to normal growth patterns during infancy and childhood? Did she experience the growth spurt typically associated with breast and pubic hair development that characterizes the onset of puberty, which usually occurs in girls between the ages ten to fourteen years? At what age did she begin to menstruate, and how often did she get her period? Menstrual cycles may be irregular at first, and some young women may go for months without a period. Irregular cycles don't necessarily predict future problems, but ongoing irregularities warrant an evaluation. The appropriate diagnostic blood tests for a variety of hormonal abnormalities in women are listed in Table 9.

I will also ask the patient whether she suffered from early onset of PMS mood changes. Such changes may predict more serious mood swings later on, particularly if there is a family history of depression or alcohol abuse, or if the woman herself has had episodes of anxiety or depression that lasted for more than a week before her period. On the other hand, a woman with a history of depression may notice a marked improvement in mood during pregnancy because of high estrogen levels. I recently met with a patient named Eve, a new mother who was almost paralyzed with depression. During the week before her period, Eve always suffered from serious depression, which would sometimes last for as long as four or five days after the flow started. "I felt terrible for more than a

TABLE 9 *Diagnostic Blood Tests*

Menopause	**Low Libido and Sexual Dysfunction**
FSH	Estradiol (estrogen)
LH	Testosterone
Estradiol (estrogen)	Free testosterone
Testosterone	Prolactin
Free testosterone	Thyroid Function Tests: T3, T4, Free T4, TSH
	Platelet MAO
Menstrual Irregularities	**PMS**
FSH	Estradiol (estrogen)
LH	Progesterone
Estradiol (estrogen)	Testosterone
Progesterone	Free testosterone
Testosterone	Platelet MAO
Free testosterone	Thyroid Function Tests: T3, T4, Free T4, TSH
Prolactin	
Thyroid Function Tests: T3, T4, Free T4, TSH	

week out of every month," she said. "Then I got pregnant. After the first few weeks of morning sickness, I felt great. The depression lifted. It was the best eight months of my life." Not surprisingly, the depression returned with a vengeance after she gave birth, as her estrogen levels rapidly spiraled downward.

Fertility problems, characterized by an inability to become pregnant or by multiple miscarriages, should also raise suspicions about hormonal abnormalities that may cause problems during perimenopause or menopause.

As I've said earlier, perimenopause most often occurs between the ages of forty-five and fifty but can begin anytime after age thirty-five or, in rare cases, earlier. The symptoms include hot flashes, night sweats, vaginal dryness, insomnia, headaches, and mood swings. Jennifer came to see me because she had just entered perimenopause and was having marked mood swings. I asked whether she or anyone in her family had a history of depression or anxiety. Jennifer nodded. "I was worried that I would have difficulty with depression during menopause because my mother and her mother both did," she said. "I just didn't think it would start so early. I'm still having periods, but I guess my estrogen levels have dropped because I'm depressed, I have hot flashes, and I can't sleep at night."

Memory loss, difficulty concentrating, insomnia, inability to make decisions, and a declining sex drive are also clues that a woman is nearing menopause. Beverly had not had a period for four months. Although she was not having hot flashes or night sweats, other symptoms pointed to the likelihood that she was entering menopause. "I can't concentrate on anything," she said glumly. "I'm taking a computer course, and I had to read and reread the manual so I could make sense of the material. Even worse, the next day, when I took the exam, I couldn't remember one word that I'd read. I got a *D* for the first time in my life."

I also ask patients about their use of prescription drugs, including antidepressants, antianxiety drugs, tranquilizers, and blood pressure medication. These drugs can diminish sex drive, disturb sleep, lower energy, and interfere with the menstrual cycle. Drinking large amounts of caffeine or alcohol, or heavy cigarette smoking may have similar effects.

Menopause is frequently associated with autoimmune thyroiditis, which often occurs in families and may result in thyroid underactivity, so I need to know if the patient has had any thyroid disease or if there's a family history of thyroid problems. Autoimmune thyroiditis may also be associated with other autoimmune diseases such as rheumatoid arthritis, adrenal failure (Addison's disease), pernicious anemia, diabetes, and hepatitis. Signs of low thyroid function include cold intolerance, dry skin, brittle hair, constipation, and lagging energy. Excessive thyroid activity may result in heat intolerance, heart palpitations, excessive perspiration, diarrhea, and tremors.

It is also critical to determine whether the patient has any history of breast, uterine, or ovarian cancer, or if there is a family history in close relatives such as a mother, sister, maternal grandmother, or aunt. Women with a history of breast or uterine cancer may not be good candidates for estrogen replacement therapy (See chapter 9 for possible exceptions to this rule.)

THE PHYSICAL EXAM: EVALUATING HORMONAL ISSUES

The next step in determining a diagnosis of hormonal problems is a physical exam. I start by palpating the thyroid gland, located below the

Adam's apple, to check for enlargement or nodular growths that may indicate a malignancy. I carefully note the patient's complexion, looking for the following conditions: dry skin may indicate a deficiency of thyroid hormone; overly oily skin and scalp hair that is thinning or balding may result from excessive testosterone stimulation; loss of hair under the arms and in the pubic area may indicate a testosterone deficiency. I also check the breasts for cysts or nodules, which may require further examination including a mammogram, ultrasound, or biopsy. I perform a pelvic examination to assess whether the tissues of the vaginal walls have atrophied, a sign of estrogen deficiency, and I check the ovaries for signs of enlargement, atrophy, or tumor. As well, I perform a series of simple neurological tests to determine whether the patient has suffered from memory loss or cognitive deficits.

HORMONAL LAB TESTING: A BRIEF GUIDE

A battery of laboratory tests is critical in arriving at the proper diagnosis of hormonal abnormalities that impinge on the brain. The list of conditions that affect the mind and may be related to hormonal abnormalities include menopause, estrogen deficiency as a result of menstrual irregularities, PMS, low libido, sexual dysfunction, depression, memory loss, cognitive dysfunction, Alzheimer's disease, and over- or underactive thy-

TABLE 10 *Diagnostic Blood Tests for Memory Loss, Alzheimer's Disease, and Depression*

Memory Loss, Alzheimer's Disease, Cognitive Function	Depression
Platelet MAO	Platelet MAO
Estradiol (estrogen)	Estradiol (estrogen)
Testosterone	Testosterone
Free testosterone	Free testosterone
Thyroid Function Tests: T3, T4, Free T4, TSH	Thyroid Function Tests: T3, T4, Free T4, TSH
Thyroid Antibodies: Thyroid peroxidase Thyroglobulin	Thyroid Antibodies: Thyroid peroxidase Thyroglobulin

roid function. Table 10 lists the diagnostic, hormonal, and MAO testing for memory loss, Alzheimer's disease, cognitive function and depression. Table 11 lists the thyroid tests that are useful in establishing various thyroid diseases. The reference values for these hormonal tests are listed in Table 12. In careful consultation with your doctor(s), you can use these values to help determine whether your blood levels of a hormone are within the normal range—a critical aspect of individualized hormone modulation.

Blood lipids, which are fatty substances in the blood, should be evaluated. Cholesterol levels increase at the time of menopause. HDL cholesterol (good cholesterol) levels decrease and LDL cholesterol (bad cholesterol) levels increase. Estrogen treatment reverses these changes, and I'll discuss this topic more fully in chapter 9.

Women who receive hormone replacement, including estrogen, should also have the thickness of their endometrium (the uterine lining) assessed on an annual basis through ultrasound. If the thickness is considered excessive, an endometrial biopsy may be necessary; many gynecologists can perform this procedure in their office. The ultrasound also allows an evaluation of the ovaries and can identify an early ovarian cancer or the inactive, atrophic ovary typical of menopause.

Osteoporosis is another condition for which all menopausal women and any perimenopausal women who have prolonged estrogen deficiency should be tested by obtaining a measure of bone mineral density (BMD). BMD is an important predictor of future fractures and can be assessed using sophisticated X-ray techniques. Excessive urinary calcium excretion can be a predictor of developing osteoporosis, as is increased urinary hydroxyproline, a biochemical indicator of bone breakdown.

TABLE 11 *Thyroid Diagnostic Blood Testing*

Hypothyroidism (underactive) and Hyperthyroidism (overactive)	Autoimmune Thyroiditis
Thyroid Function Tests:	Thyroid Antibodies:
T3 (triiodothyronine)	Thyroid peroxidase,
T4 (thyroxine)	Thyroglobulin antibody
Free T4 (free thyroxine)	Thyroid Function Tests:
TSH (thyroid stimulating hormone)	T3, T4, Free T4, TSH

TABLE 12 *Reference Values for Blood Hormone Levels*

Hormone	Female Range	Male Range
Follicle Stimulating Hormone (FSH)		
preovulatory	1.4–9.6 mIU/ml	basal level
ovulatory surge	3–100 mIU/ml	4–20 mIU/ml
postovulatory	3–10 mIU/ml	
postmenopausal	34–142 mIU/ml	
Luteinizing Hormone (LH)		
preovulatory	0.8–26 mIU/ml	basal level
ovulatory surge	25–57 mIU/ml	up to 25 mIU/ml
postovulatory	1–10 mIU/ml	
postmenopausal	40–152 mIU/ml	
Prolactin	0–20 ng/ml	0–20 ng/ml
Estradiol		
preovulatory	20–70 pg/ml	basal level
ovulatory surge	>200 pg/ml	up to 40 pg/ml
postovulatory	20–110 pg/ml	
postmenopausal	3–21 pg/ml	
Progesterone		
preovulatory	0.2–0.8 ng/ml	basal level
postovulatory	2–26 ng/ml	<2 ng/ml
Testosterone		
basal level	25–75 ng/dL	300–900 ng/dL
postmenopausal	10–20 ng/dL	
Free Testosterone		
basal level	1–2 pg/ml	20–40 pg/ml
postmenopausal	0.15–0.6 pg/ml	
Thyroid Function Tests		
Triiodothyronine (T3)	86–187 ng/dL	86–187 ng/dL
Thyroxine (T4)	4.5–12.5 mg/dL	4.5–12.5 mg/dL
Free Thyroxine (Free T4)	0.8–2.0 ng%	0.8–2.0 ng%
Thyroid Stimulating Hormone (TSH)	0.3–5.0 mIU/ml	0.3–5.0 mIU/ml
Psychiatric Testing		
Platelet Monoamine Oxidase (MAO)	1.0–2.4 nanomoles tyramine metabolized	1.4–2.3 nanomoles tyramine metabolized

ESTROGEN MODULATION FOR MOOD DISORDER

My experience in treating severely depressed women at the Worcester State Hospital made me aware that high pharmacologic doses of estrogen were required to treat major depressive disorders. These women, who had failed to respond to the conventional antidepressants that were available in the 1970s, showed clear signs of lessening depression in response to high-dose estrogen. However, in my clinical practice, I have learned that less severely depressed women will respond well to lower doses of estrogen. In some cases, estrogen therapy alone is sufficient to relieve symptoms of depression. In other cases, the lower doses of estrogen need to be combined with an antidepressant, most usually a selective serotonin reuptake inhibitor (SSRI), such as Prozac, Zoloft, or Paxil.

The estrogen dosage may have to be adjusted up or down, depending on the patient's response. At the outset, when the patient is most depressed, a higher dose of estrogen may be necessary; it can be lowered as the patient recovers and needs only maintenance therapy. Estrogen is similar in this way to other antidepressants; once the patient begins to respond positively, the high doses of an antidepressant can be reduced and finally reach maintenance levels when the person's depression is in remission.

It is important to measure blood levels of an antidepressant during treatment and adjust dosages accordingly. The same is true for estrogen. Blood levels of estrogen (estradiol) should be followed during treatment and maintained within a physiological range. (See Table 12, page 215.) Blood levels of estradiol can vary as much as tenfold across individuals receiving the same dose of estrogen. Table 13 lists the dosages of estrogen generally used for menopausal treatment compared to estrogen doses I have found effective in the treatment of depression and impaired cognitive functioning, such as memory deficit or the ability to focus, concentrate, or make decisions. I will discuss fully in chapter 9 the safety concerns of using these estrogen dosages.

NANCY'S STORY: DEPRESSION AND FATIGUE

Nancy, whom we met in chapter 4, is a good example of why it's critical to monitor blood estrogen. Nancy had a history of depression and suffered from constant fatigue, until I increased her estrogen dosage and

TABLE 13 *Standard Estrogen Doses for Menopause Treatment vs. Estrogen Doses for Treatment of Depression and Impaired Cognitive Function*

Trade Name	Generic Name	Menopause Daily Dose	Depression and Cognition Daily Dose
Oral Preparations			
Premarin	Conjugated equine estrogen (estrone sulfate and equilin sulfate)	0.625 mg	0.9–1.25 mg
Ogen	estropipate	0.625 mg	1.25 mg
Ortho-Est	estropipate	0.625 mg	1.25 mg
Estrace	17-beta-estradiol	0.5–1.0 mg	1.5–2.0 mg
Estratab	esterified estrogens	0.625 mg	2.5 mg
Transdermal Preparations			
Estraderm	17-beta-estradiol	0.05–0.1 mg	0.15–0.2 mg
Climara	17-beta-estradiol	0.05–0.1 mg	0.15–0.2 mg
Vivelle	17-beta-estradiol	0.05–0.1 mg	0.1–0.2 mg
Alora	17-beta-estradiol	0.05–0.1 mg	0.15–0.2 mg

added methyltestosterone. I had not spoken with Nancy for several months when she called to tell me that she was once again experiencing depression and fatigue. I ordered a blood test, which showed that her blood estrogen level was very low. For some reason, the oral estrogen was either not being absorbed by her body or was being metabolized too quickly. I switched her to an estrogen patch applied to the skin, and she quickly recovered. We may never know why she was not benefiting from the oral estrogen, but her case clearly demonstrates the need to monitor blood estrogen levels.

POLLY'S STORY: ANXIETY AND INSOMNIA

Another of my patients, Polly, was having frequent anxiety attacks and had been suffering for months from insomnia. Her previous doctor had put her on a standard dose of 0.5 mg oral Estrace. I gradually increased the dosage to 1.5 mg and added low-dose methyltestosterone because of her very low blood testosterone. Polly improved only slightly. Frustrated and still beset with anxiety, she made repeated calls to her gynecologist and me for advice. Her gynecologist finally decided that Polly should consult a psychiatrist about the possibility of being treated with an antidepressant and antianxiety medication.

I suggested that we check her blood estrogen level. When the results came back from the lab, I was very surprised to find that her level was extremely low. Once again, an oral estrogen was not producing adequate blood estrogen levels, even when administered in high doses. When I switched Polly to the skin patch, her anxiety disappeared, and she began sleeping through the night without interruption.

I have no real answer as to why Nancy's and Polly's symptoms were not relieved by more than adequate estrogen doses. (I do know doctors and patients alike must always consider the patient's biochemical individuality—people absorb and metabolize virtually every kind of medication somewhat differently.) In such cases, if I had not measured blood estrogen, Polly's and Nancy's problems would have remained undiagnosed, and they undoubtedly would have been referred to a psychiatrist. Physicians often overlook this laboratory test, so crucial for accurate diagnosis and monitoring for many women receiving HRT. If you feel that your symptoms are not improving as much as you would like on HRT—if you are still experiencing depression, insomnia, or anxiety—you may want to request that your blood estrogen level be tested before you begin taking an antidepressant.

INGRID'S STORY: MODULATING ESTROGEN AND THYROID

Consider the case of Ingrid, which exemplifies how hormone modulation is a process of carefully calibrating dosages until the desired results are achieved. Ingrid had been diagnosed with depression when she was in her early twenties. Over the next thirty years, she had been treated with a variety of antidepressants, which most recently included Prozac, Paxil, and Zoloft. Her PMS symptoms were quite severe and consisted of overwhelming depression, agitation, and anxiety. She had been hospitalized on five occasions, each time several days before getting her period, often because she was having suicidal thoughts.

When I first saw Ingrid, she was fifty-four and hadn't had a period for the previous six months. For two years before that, her cycle had been quite irregular. Her depression had worsened, and she was not responding to antidepressant treatment. During our first appointment, she told me that she had frequent suicidal thoughts, "but I would never actually kill myself," she assured me. Her dispirited expression and slumped shoulders underscored her words: She was clearly in great emotional pain and was feeling increasingly hopeless.

I suspected that she was in menopause, and her blood test results confirmed my belief: Her estrogen levels were extremely low and her FSH was very high. Her MAO was elevated, as well, another sign of a neurotransmitter dysfunction associated with depression. She also had a marked elevation of her thyroid-stimulating hormone (TSH), a low thyroxine level and thyroid antibodies, indicating that she had an autoimmune disorder known as Hashimoto's thyroiditis. This condition, as I discussed in chapter 2, produces a chronic inflammation of the thyroid gland; it is characterized by the presence of thyroid antibodies that attack the gland, creating deficiencies of thyroid hormone. The incidence of Hashimoto's thyroiditis increases at menopause.

I decided that both of Ingrid's hormonal problems—her estrogen deficiency and the low thyroid—should be simultaneously treated, so I prescribed estrogen and thyroid hormones. But when I saw her several weeks later, her mood had worsened rather than improved, even though her psychiatrist had also raised the dosage of her antidepressant. Another round of blood tests told me the following: Her estradiol level had increased only slightly, but the TSH level had dropped almost to a normal range, which meant that her thyroid was responding to treatment.

I increased her estrogen dosage and was delighted to hear from Ingrid three weeks later that she felt a slight improvement in her mood. I was not surprised, because her estradiol level had risen from 58 to 85 pg/ml. I wanted to see the estradiol reach the range of 125 to 150 pg/ml, so I again increased her estrogen, this time to the maximum dosage usually prescribed. One month later her estradiol level was 145 pg/ml, her MAO had decreased significantly, and her TSH was normal. Ingrid's mood happily reflected these positive changes in her blood test results. Her anxiety had subsided and her depression had disappeared, along with the suicidal fantasies.

After several months, Ingrid's mood had improved so much that I decided to see how she would react to a lesser amount of estrogen. I gradually reduced the dosage at the same time that her psychiatrist lowered the levels of her antidepressant medication. Ingrid responded positively on all counts: Her mood continued to be stable and upbeat, and the symptoms of the hypothyroidism remained in remission.

Ingrid's story is representative of what I regularly encounter in my clinical practice. Many patients initially require higher doses of estrogen before their depressive symptoms disappear. Once the remission occurs, however, doses can be gradually tapered to lower levels, as can the doses of antidepressant. Why do some women require this "kick-start" pattern of medication? Perhaps the higher estrogen doses are necessary to activate serotonin and adrenergic receptors in the brain, which have been suppressed as a result of low estrogen levels. Once the receptors are active, low doses of estrogen may be sufficient to maintain a healthy emotional state.

THE PRELUDE: PERIMENOPAUSE, MICHELLE'S STORY

When Michelle arrived for her first appointment, she brought with her a laundry list of debilitating physical and emotional symptoms. As a teenager, Michelle had devoted every free moment to riding horses, but now, one month short of her forty-sixth birthday, she was experiencing so much fatigue and muscle pain that she could hardly ride. She also suffered from crippling migraine headaches, which had first appeared shortly after she began to menstruate at age fourteen. The crippling headaches, which occurred about six times a year, "felt like an ice pick was stuck through my eye. The pain is devastating," she told me. "I'd rather have labor pains."

Michelle had three miscarriages before she was finally able to bring a pregnancy to term. She and her husband had wanted a second child, but she could not get pregnant again, even though they consulted a fertility specialist and underwent numerous hormonal treatments. In the seven years since the birth of her son, her migraines had become more severe and more frequent, and she was having trouble sleeping more than four or five hours a night. And that wasn't all, Michelle mournfully explained. During her pregnancy, she had developed pain in her lower back and hip muscles that had been diagnosed as fibromyalgia. The symptoms had gotten worse postpartum, and she was now being treated with nonsteroidal anti-inflammatory drugs, muscle relaxants, and weekly massage therapy.

For the last two years, she had also been having visual disturbances in which a portion of her visual field was blocked out for a few seconds. The first episode had occurred while she was driving along a major highway. "I had to stop the car and call my husband on the cell phone so that he could come to get me," she said. Sometimes, just before she got her period, she would notice flashing lights if she moved her head quickly and the periphery of her vision would be bordered by a crescent of sparkling zigzag lines. While migraines are often preceded by such an aura, Michelle's visual problems were not related to her headaches.

Michelle's precarious physical condition was mirrored by her fragile emotional state. She was plagued by wild mood swings and progressively worsening depression and anxiety. Her memory and ability to concentrate were also badly impaired. "I have to make lists of everything I need to do during the day, or I totally forget," she said. Her libido was declining to the point that sex had become an effort. "I'm tired of pretending I'm interested in making love. I love my husband, but it's no good to have to fake it. I can't go on like this anymore."

Although she was still having regular periods, I was quite certain that Michelle was in perimenopause, and in fact, she had low estrogen levels and a moderately elevated FSH. Her testosterone levels were all but undetectable, which fit the picture because low testosterone can be a major factor in migraine headaches. I, therefore, prescribed an oral estrogen, Premarin 0.9 mg daily, which she tolerated without difficulty. But two weeks later, she reported no change in the frequency or severity of her headaches, so I added a low dose of methyltestosterone. Michelle didn't have much faith that the combination of hormones would make any difference. She was so disheartened by her lack of improvement that, as she left my office, she said, "This is a waste of time. I would rather have M&Ms. At least they taste good."

She came back a month later, and for the first time since I'd known her, she was smiling when she walked into my office. "I haven't had a headache in four weeks, and I actually went horseback riding for two hours last weekend." Tears welled up in her eyes. "I'm having much less muscle pain, and the fatigue is gone. I never expected anything like this."

There was still more room for improvement, so I increased the dose of methyltestosterone. Over the course of the next several months, Michelle's sex drive improved, her memory and concentration returned, and her insomnia lessened. She continues to check in with me on a regular basis, because I worry that her headaches may return, even though treatment is unchanged. For the moment, however, Michelle's life has certainly changed for the better.

PREMATURE MENOPAUSE: JILL'S STORY

Jill had expected that she would enter menopause at a relatively early age because her mother's periods had stopped when she was only in her early forties. But at age thirty-three, she was not psychologically prepared for the hot flashes and missed periods that signaled the beginning of menopause. "I'm too young for this!" she protested to her gynecologist.

A daily combination of an estrogen and a progestin banished the hot flashes, and Jill experienced no other problems for five years. Then she gradually became aware that she was feeling so tired in the evenings that she could hardly wait to get into bed. She did not feel rested even after a full night's sleep, and she was crying for no apparent reason. Her gynecologist switched her to a combination of three different estrogens, taken daily, along with 100 mg of micronized progesterone. Jill did fine on this regimen for another five years, but then the depression and fatigue returned, and her sex drive faded away. Because her estrogen levels were normal, her gynecologist suggested she consult a psychiatrist. But Jill resisted the idea of taking an antidepressant and instead chose to see a psychotherapist.

Her therapist, who had referred patients to me in the past, suggested that Jill consult with me to determine whether there might be some other hormonal cause for her depression. As I routinely do with patients who complain of depression, I had her blood tested for MAO levels, which were definitely elevated. MAO elevations are often an indication of estrogen deficiency, but low sex drive is also symptomatic of testosterone deficiency, which can influence MAO. Not surprisingly, Jill's testosterone levels were way down. But she was feeling very fatigued, as well as depressed. I encouraged her to stop using the daily progesterone, and I also switched her to Premarin, a more conventional oral estrogen, rather than the combination she'd been taking. Her fatigue lessened within several weeks, as did the depression, but she still did not feel completely herself.

It was time to add low-dose testosterone to help restore Jill's nonexistent sex drive and revive her energy. I also felt that testosterone would help relieve her depressed mood. She lives on Cape Cod in Massachusetts, a considerable distance from my office, so we made an appointment to speak by phone two months later. The lilt in her voice immediately told me the story. "I'm feeling much better," she said. "My depression has completely cleared, and my sex drive is returning."

My next concern was how to protect Jill's uterus from the unopposed estrogen. Because many women develop PMS-like symptoms in reaction to oral micronized progesterone, initially I prescribed 200 mg of vaginal micronized progesterone to be used at bedtime for ten days each month. When we spoke two months later, she sounded exuberant. Her periods were normal, and she had boundless energy. The anxiety and depression were completely gone, and her sex drive was back to normal. Her improvement was nothing short of dramatic.

Jill visits me in the office once a year, and we talk on the telephone every three months. She sees her gynecologist twice a year and has an endometrial ultrasound once a year to make sure the progesterone is protecting the health of her uterus. She also has a yearly mammogram. Her primary care physician and I work together to monitor the levels of estradiol and testosterone in her blood. I also monitor her cholesterol levels to check that they are not affected by testosterone treatment. I believe that testosterone has an overall beneficial effect on heart disease in women. But methyltestosterone, which is metabolized in the liver where HDL [good] cholesterol is synthesized, can lower HDL blood levels. I will discuss this subject more fully in chapter 9.

The diagnosis of Jill's problems required an understanding of the effect that hormones can have on the mood and sex drive. Her treatment also demanded a willingness to adjust her prescriptions in order to meet her specific complaints and needs. Dosages must be changed as necessary. Side effects cannot be ignored, and innovative techniques should be employed to avoid them. This, in essence, is what hormone modulation is all about.

SURGICAL MENOPAUSE: PROBLEMS WITH MOOD, SEX, AND COGNITION, IRENE'S STORY

Irene was a talented artist who had begun drawing and painting as a young child. No matter what else might have been going on in her life, she had always depended on her creativity to nurture and inspire her. Now, however, that inspiration seemed totally to have disappeared, and she could not motivate herself to try again. When she came to see me at her first appointment, she sounded every bit as miserable as she looked. At forty-seven, she felt as if she were falling apart, and she was convinced that her problems were somehow related to the hysterectomy she'd undergone two years earlier.

Irene's gynecologist had suggested that she have the surgery because her uterus was prolapsed (dropped down) and was pressing on her bladder, causing urinary incontinence (lack of bladder control). A hysterectomy combined with a bladder repair would correct the problem. Her gynecologist had further advised her to have both ovaries removed to prevent the future possibility of ovarian cancer. Although there was no family history of ovarian cancer, Irene had gone along with his suggestions. The surgery was a success, but Irene shortly began to experience symptoms of full-blown menopause: hot flashes, lack of sex drive, insomnia and exhaustion, emotional mood swings, memory and concentration problems. Whenever she went into her studio to paint, she would soon become so unfocused and distracted that she had to walk away from her canvas.

Seeking relief from the severe anxiety and depression, she consulted a psychiatrist who prescribed several antidepressants, one of which she took in the evening to help her sleep. The insomnia was slightly improved, but Irene still felt exhausted much of the time and could not reach orgasm, a side effect of the antidepressant. Her gynecologist then prescribed various hormone combinations, including a mixture of estrogen and high doses of testosterone that he hoped might relieve her hot flashes and low sex drive. The testosterone made her feel irritable and angry, so she switched to an oral estradiol preparation, Estrace. However, even at the maximum dose of 2 mg daily, the hot flashes continued, and she was still waking up night after night, soaked with perspiration.

Irene came to see me at the suggestion of her psychiatrist. My first thought upon meeting this obviously intelligent and talented woman was that the unnecessary and unwise surgical removal of her ovaries had spun her life into a downward spiral. The ovary secretes testosterone as well as estrogen, and when the ovaries are surgically removed from a young woman whose ovaries are still actively secreting hormones, the sudden testosterone deficiency can have serious consequences. Irene's estrogen levels were in the low normal range, but her testosterone values were extremely low, and her MAO activity level was elevated.

For some women, estrogen replacement is not sufficient to control hot flashes and night sweats, but the addition of testosterone successfully relieves these debilitating symptoms. Although the high doses of testosterone had made Irene irritable and anxious and she had stopped the medication, I felt that she might respond positively to low-dose testosterone replacement. When methyltestosterone was added to Irene's estrogen, her hot flashes and night sweats began to subside. After several months, she noticed a gradual improvement in her sex drive, but she was still not experiencing orgasm.

As her mood swings stabilized, her psychiatrist lowered the dose of her antidepressant. When we talked over the next few months, I could see a real difference in her. Her memory and concentration were returning, and she had started painting again. I suggested that she speak with her psychiatrist about further reducing her antidepressant dose, which I felt might permit her to have fully normal orgasms.

Although there is a marked reduction in hot flashes and night sweats, Irene continues to have sleep difficulties. Her estrogen and testosterone levels are now within the normal range, but she still does not feel as well as she did before her surgery. Further modulation of her hormone treatment may be required to achieve that goal, and if that's not sufficient, Irene will need ongoing treatment with antidepressants or mood-stabilizing medications.

I hope Irene's experience will raise a red flag. Women should carefully consider the pluses and minuses of having their ovaries removed as a way to prevent ovarian cancer. The lifetime incidence of ovarian cancer is 1.4 percent; thus, for no good reason, ninety-eight out of one hundred women are undergoing this procedure, which can wreak havoc with their lives. However, a strong family history and positive tests for certain genes associated with ovarian cancer certainly do indicate an increased risk. If you are concerned, ask your doctor about these tests. Pelvic ultrasounds and CA-125 blood tests are screening techniques that allow for early detection of ovarian cancer. As Irene's case demonstrates, with all of our current knowledge, we still may not be able to completely reverse the hormonal deficiency that occurs following removal of the ovaries. I urge my patients—and all women—to think very carefully before deciding upon such surgery. The consequences can be devastating.

TEMPEST ONCE A MONTH: PMS, KAREN'S STORY

Karen came from a family with a strong history of depression. Her mother's sister and two cousins had been treated for depression, and Karen herself had been hospitalized while doctors searched for the right antidepressants. By the time she turned thirty-five, she had tried seven different antidepressants and three different mood stabilizers, none of which had helped her very much. Karen suffered from a type of depression known as seasonal affective disorder syndrome (SADS). She particularly dreaded the end of autumn, because she knew that her black moods would be much worse during the dark winter months. As the winter season deepens and the days shorten, our exposure to light decreases, and people with SADS become more depressed. Karen obtained some relief by exposing herself for several hours a day to a high-intensity light source, commercially available as a light box.

Karen's worst bouts of depression and anxiety occurred during the ten days before her period, when she also had hot flashes and night sweats and suffered from insomnia. She would eat uncontrollably although she wasn't really hungry and resorted to bulimia as a way to control her weight. Hers was one of the most severe cases of PMS I had ever encountered, and her chronic depression was so clearly related to her menstrual cycle that I felt that there must be a hormonal component. But why was she was so resistant to antidepressant treatment? The results of her laboratory testing gave me an important clue.

Karen's MAO levels repeatedly showed marked elevation, which was typical of my PMS patients. Earlier studies in our laboratory had demonstrated that estrogen treatment could lower high blood MAO activity and improve the mood. Karen's own blood estrogen fluctuated a great deal and was often in quite a low range for a thirty-five-year-old woman. A low estrogen level may have been contributing to her high MAO levels. But her pituitary hormones, FSH and LH, were within the premenopausal range and did not indicate an early menopause or even perimenopause.

When I first saw Karen, she was taking Paxil, an SSRI antidepressant with similarities to Prozac. Her excellent psychiatrist, Dr. Wilfrid Pilette, was familiar with our studies demonstrating that estrogen had antidepressant qualities and was comfortable with my suggestion that Karen be treated with estrogen in combination with her antidepressant. When I saw her a month later, her depression had improved somewhat, but she still was having bad days. I slightly increased the dosage of estrogen and also prescribed 200 mg of micronized progesterone to be taken at bedtime for the first ten days of each month. After several months, the PMS symptoms had almost totally disappeared, and the addition of progesterone gave Karen a regular menstrual flow for the first time in her life.

I continued treating Karen for several years, and every winter, as the days grew shorter, I had to increase her Estrace dosage to ward off depression. The higher dose resulted in some unpleasant side effects, including breast tenderness and a decrease in her sex drive. I knew that testosterone might prevent the breast tenderness and help her sex drive, but because Karen had a history of chronic cystic acne, she and I decided to undertake a trial of very low-dose methyltestosterone. We were both delighted when this turned out to be the correct choice. Her sex drive improved, and her breast tenderness disappeared.

Three years later, Karen was not having PMS or depression. Her estradiol levels remained in the normal range, and she no longer suffered from hot flashes, night sweats, or insomnia. She had stopped taking her antidepressant, and when I saw her several weeks ago, her mood was enthusiastic. She and her family were in the process of moving into a new house, which they had just built. It was the middle of winter. She had briefly increased her estrogen dosage, then decreased it when she experienced some breast tenderness. Karen has learned to adjust her estrogen dosage as her mood fluctuates. "I feel in control of myself for the first time ever," she told me.

HORMONE MODULATION: A CHOICE FOR YOU AND YOUR PHYSICIAN

As you embark on procuring hormone modulation, use the above cases as a guide, enabling you to recognize that it often takes work and time before you arrive at the right regimen. Some patients respond well to the first prescribed regimen, but others need adjustments over weeks or months before their symptoms are relieved without significant side effects. Also, for some women no regimen is absolutely perfect. A modest sacrifice may be necessary for a larger good. For instance, I have prescribed estrogen along with an antidepressant for many women with mood and cognitive impairments. When these women have sexual side effects, such as a loss of desire, the addition of testosterone frequently solves their problem. But in some cases, their libido improves but does not return to "premenopausal" levels. The upside for these women usually outweighs the downside, and while the result isn't perfect, it's far preferable to being deeply depressed and unable to concentrate. Moreover, we can continue to work on improving such women's sex drive by adjusting testosterone, and sometimes, with skillful psychotherapy.

The cases I have presented here offer you a glimpse into the art and science of hormone modulation. As you can see, it is not a simple "take this pill and call me in the morning" process. Rather, it's a "let's work together, be flexible, and make adjustments to meet your needs" process. I would argue that many fields of clinical medicine would benefit from such an approach. But in the area of hormones for women, the latest research on hormones and the brain make it imperative that clinicians and patients work together to ensure that the whole woman is helped by treatment.

9

The Safety Equation

WEIGHING RISKS AND BENEFITS

Hormone modulation may not be for everyone, but I've found that the vast majority of women with serious hormone-related disorders of mood, cognition, and sexuality can safely use individualized hormone modulation, and take rational steps to reduce the risks. Of course, *you* must make this final decision with the help of qualified medical personnel, and with careful consideration to your own medical history, family history, genetic factors, personality, and unique set of symptoms.

When making decisions about the safety of hormone replacement, you have to weigh the evidence regarding risks versus benefits. You'll need a detailed knowledge of both your own risk profile, and the specific risks associated with hormones. Since menopause has become such a focus for media stories, and the use of estrogen remains so controversial, you are likely to be overwhelmed at times by stories and claims about hormone replacement, some of which are true, some false, and some just plain confusing. I will try to help you sort the fact from the fiction.

As you prepare to make decisions about hormone therapy, consider the following list of benefits and risks associated with hormone replacement with estrogen that includes some form of progesterone. Every woman must balance this list for herself, given her own history and concerns. This first list conforms to today's conventional understanding:

Risks (increased incidence)

Breast cancer, after long-term use (controversial)
Thrombophlebitis (blood clot in leg)

Pulmonary embolism
Breast tenderness ("mastalgia")

Benefits (decreased incidence and/or severity)

Hot flashes/night sweats/insomnia
Heart disease: incidence and deaths
Osteoporosis
Stroke
Macular degeneration
Fibromyalgia
Colon cancer
Overall cancer deaths
Deaths from all causes

Among the risks, it should be noted that the breast cancer connection remains controversial, but even if you accept the results of a recent large epidemiological study, an increased risk is evident only after ten years, and the increased mortality after that time is modest. The increased incidence of thrombophlebitis—clots in the deep veins of the leg—is small, and pulmonary embolism is quite rare. Meanwhile, the risk of death from heart disease—the biggest killer of women—is lowered roughly 40 to 50 percent. Moreover, according to that same large study, after one decade on hormone replacement the risk of death from *any* cause is cut down by 37 percent.

The above-listed benefits refer to those proved in clinical research. However, in my clinical experience certain other conditions may also be prevented or treated by hormone therapy including estrogen. Among these conditions are arthritis, migraine headaches, and irritable bowel syndrome (IBS). The medical literature includes conflicting reports on arthritis and migraines, and no significant data on IBS, but I have had a number of patients with these conditions who experienced considerable improvement once they began hormone treatment.

Now let's look at the list of benefits again. This time, however, I will add the benefits of hormone therapies to the brain—the ones that have been documented over the past several decades. Please note that the psychological and cognitive disorders listed refer only to those associated with hormonal decline during or after menopause:

Benefits (decreased incidence)

Hot flashes/night sweats/insomnia
Heart disease: incidence and deaths

Osteoporosis
Stroke
Macular degeneration
Fibromyalgia
Colon cancer
Overall cancer deaths
Deaths from all causes
Mood disorders
Treatment-resistant depression
Cognitive impairments
—Short- and long-term memory loss
—Fuzzy thinking
—Difficulties with concentration and focus
—Difficulties with decision-making
—Verbal slips
Alzheimer's disease
Sexual disinterest
Sexual dysfunction

Peri- and postmenopausal women ought to consider this expanded list of benefits, and weigh them against the risks as they consider hormone treatment.

By presenting this expanded benefits list, I do not mean to suggest that the hormone decision should be a simple cut-and-dried matter for every woman. I will never minimize any woman's concern about breast cancer—the only risk that greatly concerns most women. An increased incidence of uterine cancer is no longer a concern once progesterone is properly added to hormone replacement. But I believe that women must put the reported breast cancer risk into proper perspective.

HORMONE THERAPY AND BREAST CANCER: A BALANCED VIEW

Many women fear that estrogen will raise their risk for breast cancer. Is this dread justified? The jury is still deliberating, despite more than fifty scientific studies seeking to establish a relationship between estrogen use and breast cancer. The majority of these studies have demonstrated no such relationship.

Several large "meta analyses"—statistical overviews of many studies—have shown no connection between estrogen use and breast cancer. A few have shown a very small increased risk after many years of use. For example, a major analysis of fifty-one epidemiological studies including over 160,000 women indicated a very modest increased risk of breast cancer after many years of use. The researchers used their data to make these predictions: Among women between fifty and seventy who never use estrogen, about 45 out of 1,000 will develop breast cancer. Among those who use estrogen for five years, the number rises to 47 out of 1,000. At ten years, it's 51 out of 1,000, and at fifteen years, it's 57 out of 1,000. Put another way, 1.2 percent of those 1,000 women will develop breast cancer as a result of fifteen years of use. Thus, if estrogen does cause an increased risk of breast cancer, it is likely to be very small. Moreover, the types of breast cancer most strongly related to estrogen are probably the most easily cured—a point to which I will return.

The Brigham and Women's Study

One of the more important and frequently quoted studies on the estrogen/breast cancer issue was published in 1997 in the *New England Journal of Medicine.* Led by epidemiologist Francine Grodstein of Brigham and Women's Hospital in Boston, the study followed 60,000 postmenopausal women for sixteen years. Grodstein and her colleagues looked at the differences between women who were taking hormone replacement versus those who were not. Strikingly, the overall mortality rate was 37 percent lower among women on hormone therapy. These women were much less likely to die of *any* cause.

What about breast cancer? The risk of dying from breast cancer was actually lower among women taking hormones for ten years or less—24 percent lower. (See Figure 9.) However, after ten years of hormone use, breast cancer mortality rose by 43 percent. Consider, though, what this means in real terms. It's been estimated that about 4 out of 100 postmenopausal women will get breast cancer, and based on this one study, with hormone use the number rises to about 6. That's 2 additional women out of every 100.

While the increase in breast cancer use after ten years did influence the overall mortality statistics, overall *women who took hormones were still less likely to die from any cause.* The 37 percent cut in mortality at five years did drop, but the hormone-using women were 20 percent less likely to die than nonusers.

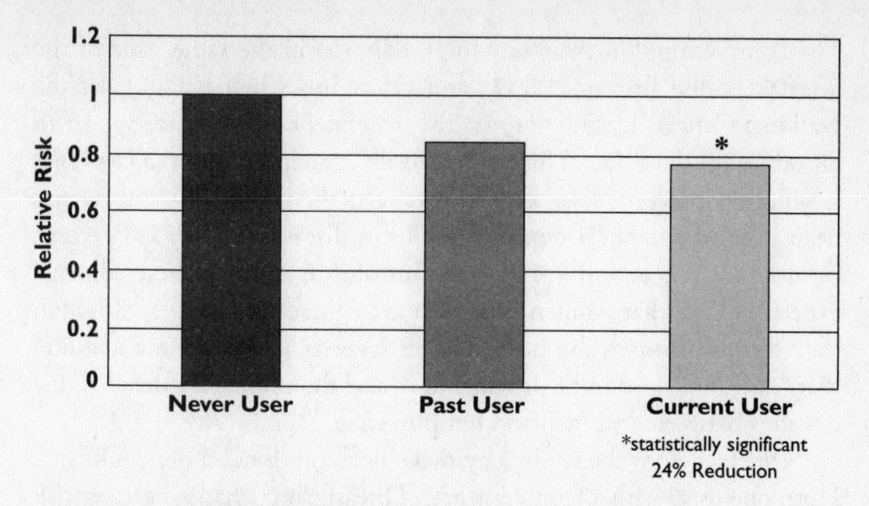

FIGURE 9 Risk of Death from Breast Cancer Among Estrogen Users
For women taking estrogen for ten years or less at the time of breast cancer diagnosis (current use), the risk of dying from breast cancer was actually 24 percent lower than that of former or nonusers. (*Grodstein, F., et al.* N Engl J Med.*1997; 336: 1769.*)

The biggest reason why hormones cut the death rate was fewer deaths from coronary artery (artery to the heart) disease. Women on hormones for ten years experience an impressive 53 percent reduction in the risk of death from heart disease. Consider these estimates, based on Grodstein's study, of risk reduction resulting from ten years of hormone replacement therapy:

Causes of Death	Estimated *Reduction* in Risk of Death
All causes	37%
Heart disease	53%
Stroke	32%
All cancers	29%
Breast cancer	24%

The findings of this study may help women recognize the overwhelming benefits of hormone use to long-term health. If there is a small increased risk of breast cancer after ten years of use, it needs to be put in perspective. It would be a modest increase, as noted above.

Commenting on the study in an editorial in the same issue of the journal, Louise Brinton, Ph.D., and Catherine Schairer, Ph.D., of the National Cancer Institute wrote, "the benefits of estrogen use appear to far outweigh the risks." This was especially true in light of the fact "that a white woman's . . . risk of death between the ages of 50 and 94 has been estimated to be 31 percent from heart disease and only 2.8 percent from breast cancer and 2.8 percent from hip fracture." These editorial experts believe that women must factor in these percentages, although they did not dismiss the breast cancer issue as irrelevant, maintaining that "decisions need to be personal ones and should involve detailed discussions between a woman and her physician."

One fact from the study may make decisions a tad more confusing. Hormone users with a family history of breast cancer had no greater risk of death than hormone users without any such family history. However, Brinton and Schairer point out that women and their doctors can consider a range of breast cancer risks—not just family history—when they make the hormone decision. (Reproductive behavior, benign breast disease, and other factors may be relevant.)

When comparing their concerns about heart disease versus breast cancer, women should take into account a few additional facts. As the *New York Times* health correspondent, Jane Brody, framed the issue, only 3.8 percent of American women who live to ninety will get breast cancer, while 50 percent will die of heart disease. Hormone replacement can cut that heart disease risk by 40 to 50 percent. Dr. Francine Grodstein makes another critical point: Of course, the breast cancer survival rate is highest when women are vigilant about early detection, with breast self-exams and regular mammograms.

What If I Can't Decide?

Those of you with reasons to be highly concerned about breast cancer, but with serious menopausal symptoms that compromise your quality of life, may want to ask yourselves several questions. The first question is, having once started hormone treatment, how long should it be continued? The answer is not obvious. The second question is, if you decide to stop hormone therapy, what are the consequences? The answer to this one is quite well known. By stopping, you may lose some of the benefits to the heart, bones, and brain. Moreover, if you stop hormone therapy, your symptoms may return with a vengeance, but you always have the option of resuming treatment, and perhaps intensifying your early detection strategies in order to ease your concerns about breast cancer. Only you can decide.

Does Estrogen Cause Less Risky Breast Cancers?

We now have striking evidence that if estrogen does increase the risk of breast cancer, it is predominantly the less common types, which are much more treatable and curable than the more common types. In 1999, Susan Gapstur, Ph.D., and Monica Morrow, M.D., of the Northwestern University Medical School, and Thomas Sellers, Ph.D., of the Mayo Clinic Cancer Center, published in *The Journal of the American Medical Association* a major study of hormone replacement involving 37,105 women.[1] They classified breast cancers according to their pathological characteristics as observed under a microscope. The researchers reported that there is little evidence to link estrogen replacement with the most common types of breast cancers, which make up 85 to 90 percent of all invasive breast cancers. (e.g., invasive, ductal, and lobular types). They did find a small increased risk of other types of less common breast cancer in women who took estrogen. These less risky types are classified as medullary, papillary, tubular, and mucinous tumors. Each of these types has a relatively high cure rate.

"These cancers have a better outlook, respond well to treatment, and are less likely to spread elsewhere in the body," said Dr. Gapstur. Dr. Trudy Bush, a leading epidemiologist at the University of Maryland School of Medicine in Baltimore, wrote an editorial accompanying the journal article. According to Bush, the study's most important conclusion was the lack of connection between HRT and the most common forms of breast cancer. In a telephone interview with the *New York Times,* Dr. Bush said, "We'd been looking for a connection for fifty years and it doesn't show up. Overall, there's nothing here."

This study may make it easier for menopausal women to make informed decisions regarding hormone replacement. If HRT use selectively increases the occurrence of less commonly occurring tumors that have a good prognosis, then menopausal women and their doctors should factor this into their risk-versus-benefit calculations. Women should obviously be concerned about preventing diseases that are more likely to be lethal.

BREAST CANCER AND PROGESTERONE

A recent study has suggested that estrogen use is associated with a slightly increased risk of breast cancer and that the addition of progestin to the treatment may increase the risk further. In this study from the National Cancer Institute, there was an association between the breast cancer risk and body weight. Heavier women did not have an increase in breast cancer risk with the use of estrogen alone or estrogen plus progestin whereas thinner women had a very slight increase. This increase in risk is similar to a study we discussed earlier in this chapter. Without estrogen, 4 women out of 100 developed breast cancer. On estrogen alone it would increase to 5 and on estrogen plus progestin to 6. These results differ from another large survey, which failed to identify any increased risk of breast cancer among estrogen users.[2] An editorial in *Journal Watch,* published by the Massachusetts Medical Society, stated that, "Such disparate findings suggest that any increased risk must be small and that hidden confounding factors exist in these studies."

Commenting on this new study the American College of Obstetricians and Gynecologists (ACOG) does not recommend a change in the current clinical practice of HRT administration. Their view is that additional research is needed to better define the balance of risks and benefits of HRT use.

Putting Breast Cancer in Perspective

None of my analysis is meant to suggest that women should forget about breast cancer as a concern. Women should and must do everything possible to prevent cancer with their decisions about lifestyle, early diagnosis, and hormones. Health experts in many fields are telling us that the media has induced a kind of hysteria about breast cancer, one that should be replaced by thoughtful concern and careful action. When women make the hormone decision, they must put their breast cancer concerns on one hand, and balance them against the benefits they would accrue on the other hand. Women with sound reasons to be very worried about breast cancer may indeed choose to forgo hormones. Others, with pressing symptoms today and concerns about other diseases tomorrow, may choose to initiate hormone treatment.

Can estrogen cause cancer, or can it promote a malignancy that would have shown up anyway? No one is completely certain. Organizations that raise money for breast cancer research like to say that one in nine American women develop breast cancer. However, that figure is relevant only if you happen to be eighty-five years old. This statistic has scared the wits out of women in a way that is not justified.

The incidence of breast cancer increases dramatically with age, being highest in the older age ranges when estrogen levels are lowest, and lowest in the younger age brackets when estrogen levels are highest. (See Figure 10.) Estrogen may not be causing breast cancers, but rather stimulating an already existing cancer. Some cancer biologists believe that estrogen acts as a late-stage growth promoter, and that withdrawing the hormone once cancer is diagnosed is beneficial. Further, according to one 1995 study, women on HRT for five years or less who stop treatment have a slightly reduced incidence of breast cancer. Strangely, however, estrogen is no cancer promoter when it comes to other types of malignancy. Remember, in the 1997 study by Grodstein, hormone users had a reduced risk of all other cancers. (Shortly, I'll describe how estrogen reduces the risk of colon cancer.)

More definitive answers regarding the relationship of estrogen treatment and breast cancer will be come in the future, as we learn the results of the Women's Health Initiative Study. But we won't have final data from this placebo-controlled clincal trial involving 27,000 women until at least 2005. Until then, you can use the data presented here, along with your own research and discussions with your physicians, to come to the most self-protective, health-promoting decisions about hormone treatment.

HORMONAL PREVENTION OF COLON CANCER

Colon cancer rates start to rise at age forty and peak at age sixty to seventy-five. Every year, about 50,000 Americans are diagnosed with colon cancer, and 24,000 die from the disease. Colon cancer is the third leading cause of cancer deaths in women in that age range.

Convincing data from about twenty studies reveal that estrogen reduces the incidence of colon cancer in women who have taken hormones for a period of time, though the biggest effect is on current users. In one large study of 422,373 postmenopausal women, current use of estrogen reduced the incidence of fatal colon cancer by approximately

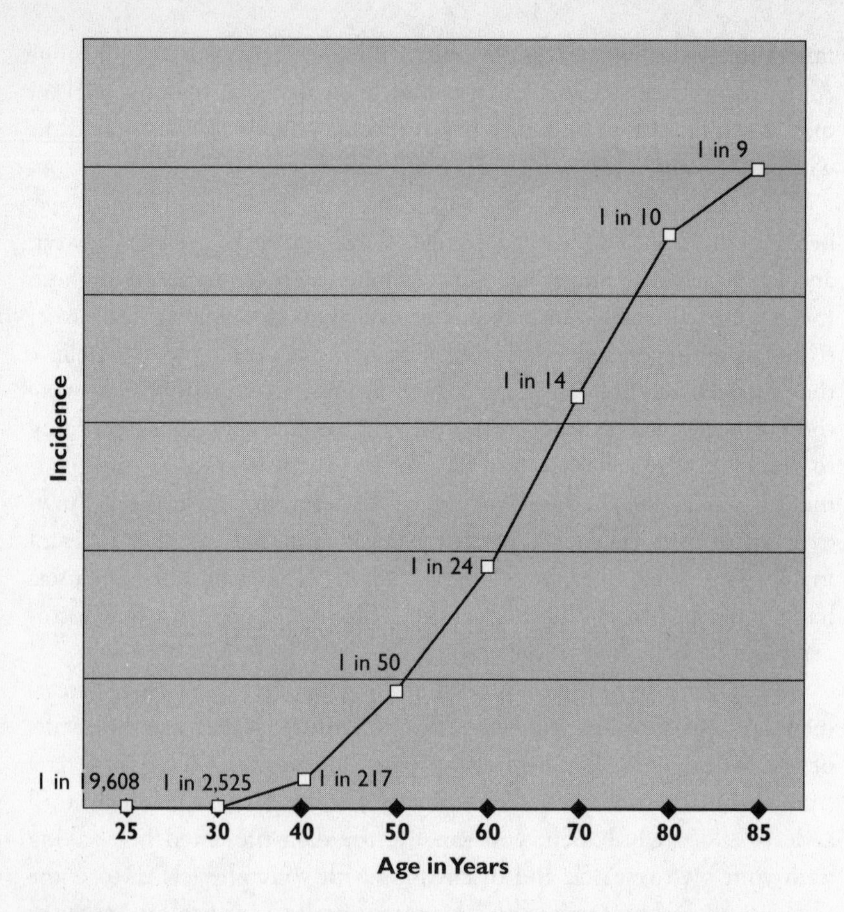

FIGURE 10 Incidence of Breast Cancer Increases with Age
The incidence of breast cancer increases dramatically with age. Data from: Lawrence, M., "The Great Estrogen Debate." (*In* Menopause and Madness, *Andrews McMeel Publishing. Kansas City, Missouri [1998].*)

45 percent. Hormonal use for one year or less resulted in a 19 percent reduction, whereas users who had been on treatment for eleven years or more benefited with a 46 percent reduction.[3] The longer a woman is on estrogen, the greater the protective effect. Just how estrogen exerts its protective effect is unknown. We do know that estrogen stimulates the immune system, which may recognize and destroy tumor cells as if they were foreign substances. Also, estrogen deficiency during menopause is associated with an increase in an inflammatory chemical released during

an acute inflammatory response. This inflammatory chemical may contribute to the development of several diseases of later life, including lymphoma (a type of cancer), osteoporosis, and Alzheimer's disease.

Theoretically, a decreased immunity would fail to recognize the specific proteins of cancer cells as foreign substances, which would be destroyed if the person were not immune deficient.

Women with a family history of familial polyposis, an inherited tendency to develop polyps in the colon, which greatly increases the risk of colon cancer, may want to consider hormone treatment as one form of cancer prevention. Indeed, any woman with a family history of colon cancer, or any other reason to suspect that they're at risk for the disease, will have another benefit—this one a potential lifesaver—from hormone therapy.

BENEFITS OF HORMONES FOR HEART DISEASE

In the last chapter, I discussed the hormonal prevention of heart disease, but I will add salient details here, since any woman's risk/benefit calculation must include heart disease.

In our cancer-phobic culture, women are rarely helped to understand that heart disease takes a much greater toll on women's lives after menopause than breast cancer. As can be seen in Figure 11 the mortality from heart disease increases dramatically more than deaths from breast cancer as women age. In women under forty-five, heart disease causes 29 percent more deaths than does breast cancer. (See Figure 12.) In women over forty-five, heart disease kills over 800 percent as many women as breast cancer! (See Figure 13.) Of course, hormone modulation is most beneficial during these peri- and postmenopausal years. Since 1985, eleven studies have consistently shown a 40 to 50 percent reduction in heart disease in women on HRT.[4]

While breast cancer risk must be considered, heart disease is far more lethal for most women in their perimenopausal or menopausal years. Figures 12 and 13 depict the various mortality rates of heart disease, breast cancer, and other causes of mortality in women.

The mechanisms by which estrogen protects against heart disease are not completely understood. As is well known, estrogen lowers total

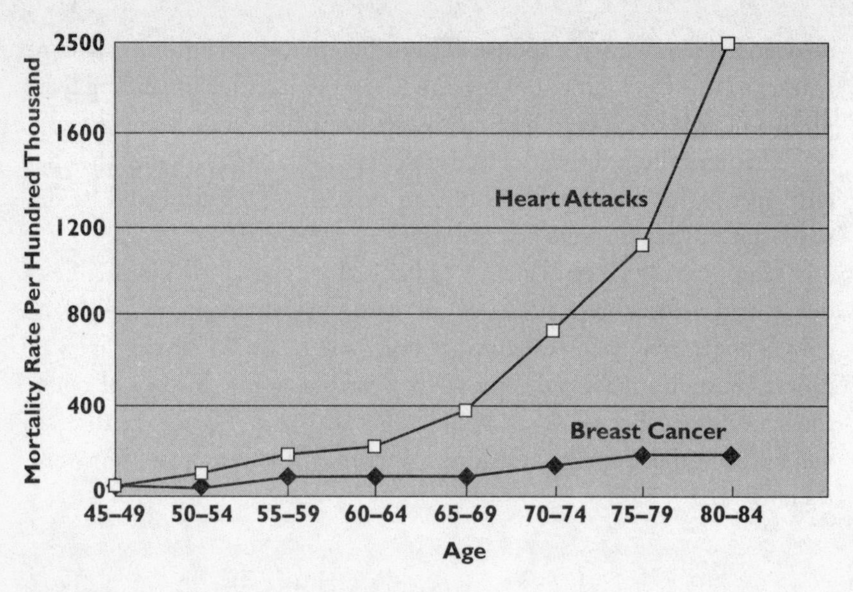

FIGURE 11 Mortality Rates in Women
Mortality rates in menopausal women show overwhelmingly that heart disease, not cancer, is the leading cause of mortality in postmenopausal women. One in two women will die of heart disease or stroke, but one in twenty-five will die of breast cancer. National Center for Health Statistics. *Vital Statistics of the United States*. 1992. *Vol. II—Mortality*, Part A. (See Cancer Statistics Review *1973–1993. Miller et al., eds. National Cancer Institute 1997.)*

cholesterol, increases HDL (good) cholesterol, and decreases LDL (bad) cholesterol. But these cholesterol effects account for only 25 percent of the beneficial actions of estrogen on heart disease. The other 75 percent of estrogen's beneficial effects include:

- *Increased blood flow to the heart*
- *Anticlotting effect*
- *Decreased deposits of LDL (bad cholesterol) in the arteries to the heart*
- *Strengthens the endothelial lining of the arteries of the heart protecting them against damage by smoking and bacteria*

These mechanisms may help us understand how, in the 1997 *New England Journal of Medicine* study, deaths from heart disease were cut by 53 percent in women using hormones for a decade. The study sug-

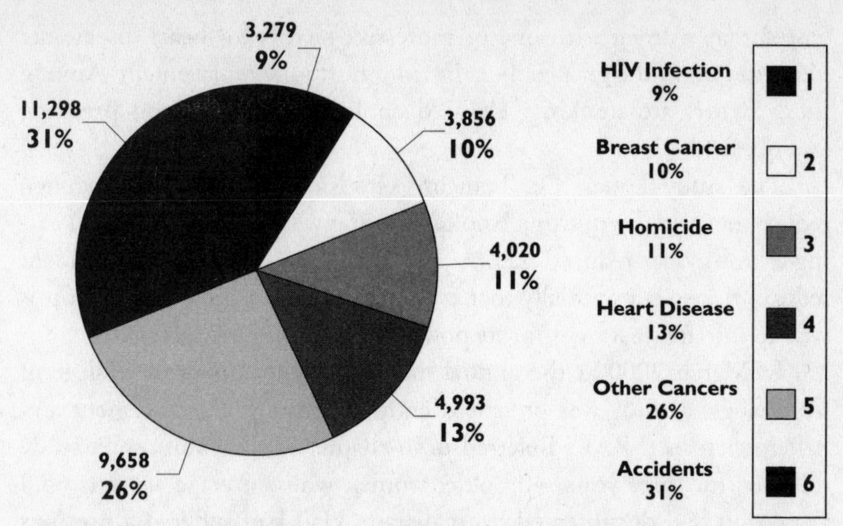

FIGURE 12 Leading Causes of Death Among Women Forty-four Years Old and Under/1992

In women under forty-five, heart disease causes 29 percent more deaths than does breast cancer.

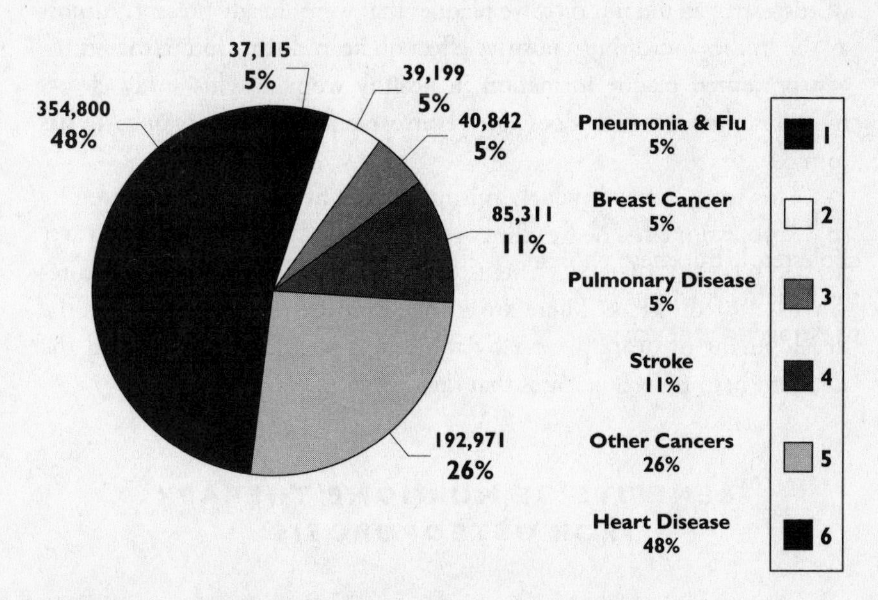

FIGURE 13 Leading Causes of Death Among Women Forty-four Years Old and Over/1992

In women over forty-five, heart disease kills over 800 percent as many women as breast cancer.

gested that women with one or more risk factors for heart disease are likely to obtain the greatest benefit from hormone replacement. Among these factors are smoking, elevated cholesterol, high blood pressure, and obesity.[5]

The study leader, Dr. Francine Grodstein, suggests that women reduce their risk by quitting smoking, losing weight, exercising, and eating a healthy, fat-reduced diet. At the same time, hormone replacement reduces risk so substantially that every woman must decide whether it is wise to initiate treatment for its powerful heart-protective benefits.

In March 2000 at the annual meeting of the American College of Cardiology, a study was presented entitled "Estrogen Replacement and Atherosclerosis (ERA)." Referred to in chapter 7, this multicenter study followed for three years, 309 older women whose average age was 65.8 years who had documented heart disease. Half had suffered a previous heart attack. The women were treated with either Premarin (conjugated estrogen), Prempro (Premarin plus a progestin—medroxyprogesterone) or placebo. The researchers studied the arteries of the heart and found that these older women who had established atherosclerotic plaques blocking the arteries were not affected by the administration of hormones. In other words, estrogen did not dissolve plaques that were already present. Almost all the studies indicating a positive effect on heart disease found that estrogen prevented plaque formation in healthy women. This study clearly indicates that estrogen does not reverse plaque formation once it has formed.

This study is being widely misinterpreted by the media and even by some professionals. The headlines read, ESTROGEN DOES NOT PROTECT AGAINST HEART DISEASE. This is totally misleading and is a misrepresentation of the facts. There are strong data that estrogen prevents the development of heart disease by inhibiting plaque formation and the ERA study in no way negates that data.

BENEFITS OF HORMONE THERAPY FOR OSTEOPOROSIS

Osteoporosis or thinning of the bones is a condition common in menopausal women that is known to be associated with estrogen deficiency and is preventable with the administration of estrogen. The extent of bone loss can be quantified by measuring bone mineral density

(BMD), which is an assessment of the calcium content of bone. The bones of women with osteoporosis are seriously lacking in calcium, reflected by a low BMD that makes them much more prone to fractures. Two recent studies have reported an association of low BMD with brain changes that may reflect estrogen deficiency.

- Women with osteoporosis have a significantly increased incidence of depression and higher levels of depressive symptoms.[6]
- In a study of 8,333 women aged sixty-five or older reported in the *Journal of American Geriatrics Society,* researchers followed the women for four to six years.
- The women whose bone density were within the top 20 percent performed as much as 8 percent better on tasks of cognitive function.[7]

These two studies suggest that estrogen deficiency may not only have a deleterious effect on bone but also on brain.

Osteoporosis currently affects more than 25 million women, resulting in 150,000 hip fractures annually. Although menopausal women are at major risk of developing osteoporosis, 75 percent have never discussed the problem with their doctors. This is a major tragedy, since this extremely debilitating disease can be prevented by hormone modulation.

Using the results of BMD, it is possible to evaluate an increased risk of fracture with minimal trauma. BMD can warn the doctor and his patient that the risk of hip fracture has doubled because of the degree of osteoporosis. Fractures can occur in the hip, pelvis, wrist, and spine. Very common results of osteoporosis are compression fractures of the spine, in which the spinal vertebrae collapse as a result of weakness. This leads to shortening and finally curvature of the spine or the so-called dowager's hump frequently seen in elderly women. By age sixty-five, 15 percent of women have vertebral fractures and by age eighty-five that figure has risen to 30 to 40 percent.

Hip fracture can be a disastrous event in an elderly woman's life, and it's the most common fracture seen with osteoporosis. In addition to the suffering and discomfort, the average cost of a hip fracture in the United States can reach as high as $35,000 per patient. The most tragic statistic: Approximately 20 percent of the patients die within the first month of injury. Most of the deaths from hip fracture occur within the first four months and are secondary to heart failure, pulmonary embolism (blood clots traveling to the lung), and pneumonia. Numerous studies have proven that HRT reduces fracture risk by 20 to 60 percent. Delaying the start of HRT for six years can result in a 10 percent reduction in bone density.

In addition to estrogen, there are a number of other alternatives for women with osteoporosis. Alendronate (Fosamax), a bisphosphonate, taken orally, has been approved for the treatment of osteoporosis. Fosamax produces a 3.5 percent increase in bone density in the spine, and a 1.9 percent increase in the hip within two years. Within 3 years 8 to 10 percent increases in spine density were observed.[8] I have observed as much as a 10 percent increase in bone density in menopausal women when hormone replacement is combined with alendronate. A study recently published suggests that there is an additive effect on bone density when estrogen and alendronate are given in combination.[9]

Calcitonin, a synthetically made medication similar in structure to the parathyroid hormone calcitonin extracted from salmon, has been approved for use in the treatment of osteoporosis. It is marketed as a nasal spray, Miacalcin, which can prevent bone loss, but is less effective than alendronate. I have successfully treated my osteoporosis patients with a combination of estrogen and Miacalcin. The effect of the two together appears to be additive. Miacalcin is particularly useful in elderly patients who are on multiple medications, since it has few negative drug interactions. Miacalcin also can be useful in cases where Fosamax has produced serious side effects, such as irritation of the esophagus, a painful problem that in rare instances can cause strictures.

HOLLY DANCES AGAIN

I'm continually impressed by the number of different ways that menopause can begin. Each woman has her own set of symptoms, unique to her own story. The traditional symptoms of hot flashes, night sweats, and vaginal dryness frequently are not among the chief complaints of many menopausal women.

Holly's story was definitely her own. Her periods had started at age thirteen but were not the usual twenty-eight-day cycle. Instead they occurred every twenty-one days. The amount of menstrual flow had been normal and was associated with moderate cramping. Holly had never had any difficulty with PMS. She had chosen not to become pregnant and pursued her career as an executive in an insurance firm. At age thirty-eight on a routine pelvic examine her doctor told Holly that she had an ovarian mass. He was concerned about the possibility of ovarian cancer, but when surgery was performed the ovarian tumor proved to be benign. About one year before the surgery a rather unusual symptom had appeared. Holly began to have painful uterine cramping every time she became sexually aroused. This discomfort was further aggravated at the time of orgasm. Immediately after surgery the cramping had markedly decreased and but continued to occur with orgasm.

When I saw her at age forty-six she had had a menstrual flow only twice a year for the last two years. Also her menstrual flow had decreased from five to seven days, so now it only lasted two to three days. She had none of the classical symptoms, no hot flashes, night sweats, or vaginal dryness. Preoperatively, before the surgery for her ovarian tumor, Holly had her pubic hair shaved and following surgery the hair did not grow back. Also her sex drive had declined in the past two years, and even though she was able to have an orgasm she avoided becoming aroused because it produced pain and cramping. She had gained fifty pounds in the previous eight years and while most of the time she slept normally, she was beginning to have episodes where she would wake up early in the morning and not be able to fall back asleep. Her energy level had declined seriously and she rarely exercised.

As if these symptoms were not enough, Holly had just recently suffered a fracture of her right leg and at the time I saw her she was in a cast from her hip down to her toes. A measure of her bone density had revealed that Holly was suffering from serious osteoporosis or thinning of the bones, which had put her at risk for bone fractures. In addition to all these physical symptoms, Holly, who had always been an optimist, was having chronic problems with a depressed mood and she found herself weeping frequently.

It was very clear to me from these symptoms that Holly was entering menopause. Her low estrogen levels were insufficient to protect her bones from the serious loss of calcium that occurs in many women as they pass into menopause. But it was not just her bones that were suffering; her brain function was being seriously impaired as a result of the hormone deficiency. This was affecting her sleep, her mood, and her energy.

Also estrogen was not the only hormone that was deficient. The decline in the other ovarian hormone, testosterone, had led to a loss of pubic hair growth, a decline in sex drive, and a debilitating loss of muscle. This loss of muscle was playing a major role in her fifty-pound weight gain. When testosterone levels drop, the muscles, which require testosterone stimulation, disappear and frequently are replaced by fat. A pound of fat burns one-seventh the number of calories as compared to a pound of muscle. This shift in body composition results in a major decrease in the body's calorie expenditure predisposing to serious weight gain. The loss of muscle lowers endurance as well.

Testosterone also is important in maintaining normal sex drive and function. Testosterone, like estrogen, plays an important role in brain function and in its absence an overall feeling of well-being is lost.

Holly's laboratory testing revealed the classical findings of menopause. A high blood level of FSH clearly identified her menopausal state. Further, her blood estrogen and testosterone levels were extremely low, also classical signs. It was absolutely clear that what Holly required was treatment with estrogen and testosterone.

However, because of her serious problem Holly required a more aggressive treatment. In addition to her estrogen I prescribed a drug that would stimulate bone growth. The estrogen treatment would prevent further bone loss and if Holly maintained a good calcium intake her condition would not worsen. If she remained as she was, she would still be prone to frequent bone fractures. What was needed was treatment that would increase Holly's bone density. A drug that is useful in this regard is Fosamax. In other patients with low bone density Fosamax had restored their bones to a close to normal state, protecting them against fractures.

Two months after I started Holly on treatment her moods had improved dramatically. Her sex drive was again vibrant and she no longer had uterine cramping and pain when she became aroused. Her sleep had returned to normal. The pubic hair had begun to grow again, and her strong endurance was restored.

A year later when we measured her bone density, it had improved by 8 percent, a very significant positive change. Just recently, four years after her initial visit, I saw her in my office. She was glowing and obviously full of energy. While she had been in physical therapy, following her leg fracture, her physical therapist had asked her, "What is your goal in recovery?" Holly told the therapist she wanted to start flamenco dancing, a fiery Spanish Gypsy dance. As she strode out of my office, she stopped at the door and turned back to me. "I'm a flamenco dancer now."

FIBROMYALGIA: A PAINFUL LACK OF HORMONES?

Fibromyalgia is a condition characterized by aching pains in many muscles in the body, in such areas as the arms, legs, back, and hips. Some experts believe that fibromyalgia is an autoimmune inflammatory disorder, but its nature remains rather stubbornly ambiguous. The disorder affects many menopausal women, and often, it's clearly related to a deficiency of estrogen and testosterone during and after menopause. It frequently occurs in women who are also having difficulty with depression, anxiety, insomnia, headaches, and irritable bowel syndrome. Like irritable bowel syndrome and some types of migraines, I consider fibromyalgia to be one of a group of diseases that are offshoots of depression. A number of studies have demonstrated that hormone replacement in menopausal women has been beneficial in the treatment of fibromyalgia. If you are peri- or postmenopausal, and you have fibromyalgia that does not respond to any other treatment, you may wish to consider hormone modulation.

HORMONES AND OSTEOARTHRITIS

There are conflicting reports about whether estrogen plays a significant role in osteoarthritis. It's been confirmed that estrogen receptors are present in human articular cartilage. It may be that estrogen modulates local inflammation in various joint diseases. A number of investigators have demonstrated a positive impact of HRT on the risk of osteoarthritis, but others have found no relationship or suggested even a negative impact from estrogen. A 1998 report from the Boston University School of Medicine summarizes the epidemiological studies on this question, and it concludes that women on estrogen replacement therapy (ERT) show a lower prevalence of osteoarthritis than women not on estrogen.

HORMONES AND MACULAR DEGENERATION

Age-related macular degeneration is the leading cause of legal blindness in the United States, accounting for 60 percent of all new cases. Between the ages of forty-three and eighty-six, 17 percent of the population is afflicted; over seventy-five years old, 35 percent develop the condition. Women with early-onset menopause have a 90 percent increased risk of age-related macular degeneration (AMD). Unfortunately, there is no medical treatment for macular degeneration, and surgical treatment is effective only in a small number of patients. But there's good news: Data from one large study strongly suggest that HRT reduces the risk of developing AMD. There is approximately a 1 to 4 percent reduction in AMD per year of HRT treatment.

LESSER RISKS: THROMBOPHLEBITIS, EMBOLISM, AND GALLBLADDER DISEASE

Thrombophlebitis is the formation of a clot in a large vein, usually in the leg. The vein can become inflamed and irritated, producing a warm tender leg calf. With oral contraceptives containing large amounts of estrogen as well as a progestin, there has been an increased incidence in thrombophlebitis, particularly in women who smoke. Two recent studies have found a slight increased risk of thrombophlebitis in current HRT users, while a third found no increased risk.

Pulmonary embolism occurs when a blood clot forms in a large vein, often in the leg or heart. These clots may not be firmly attached to the vein wall and can break off and travel in the bloodstream to the lung, producing a pulmonary embolus. Conditions such as trauma, surgery, immobilization, or cancer will increase the risk of pulmonary embolism. Among healthy postmenopausal women age fifty to fifty-nine on ERT, there is only a very slight increased risk of pulmonary embolism.

The incidence of gallstones and gallbladder disease may be slightly increased in women on HRT. Some studies suggest no relationship between HRT and gallbladder disease.

While it is important to be aware of these risks of HRT, I do not wish to frighten women who are trying to reach a decision as to whether to start taking estrogen. These particular risks are minimal, and they are clearly outweighed by the benefits of estrogen on heart disease, osteoporosis, and Alzheimer's disease.

HIGHER-DOSE ESTROGEN FOR MOOD AND COGNITION

In my practice of hormone modulation, treating depression, anxiety, memory loss, and cognitive dysfunction usually requires a somewhat higher dose of estrogen than the treatment of hot flashes, night sweats, heart disease and osteoporosis. For example, while 0.625 mg of Premarin daily may be sufficient to get the benefits of estrogen for heart disease and osteoporosis, treating mood disorders, memory loss, or cognitive impairment frequently requires doses of 0.9 to 1.25 mg of Premarin.

There is no indication that doses in this range increase the risk of breast cancer. One of the most intriguing clinical studies of HRT's risks and benefits helps to confirm the relative safety of larger-than-standard doses of estrogen. A renowned reproductive endocrinologist, Dr. Lila E. Nachtigall, Professor of Obstetrics and Gynecology at New York University School of Medicine, led this double-blind study. Half of the women received 2.5 mg of Premarin (four times the current standard dose) along with synthetic progestins seven days each month, while the other half received placebos. After ten years, 4 of the 84 women taking a placebo had developed breast cancer—4.7 percent—while none—0 percent—of the 84 women on high-dose estrogen developed breast cancer.

After Dr. Nachtigall's decade-long study was completed, patients were given the option of starting, stopping, or continuing HRT, this time with the standard 0.625 mg of Premarin and ten days of progestins. In 1992, she published her data after twenty-two years of follow-up. Once again, patients taking estrogen had no increased risk of cancer. *None of those receiving HRT (0 out of 116 patients) developed breast cancer, while 11.5 percent who never used estrogen (6 out of 52) developed the disease.* These remarkable findings from a randomized clinical study should reassure women who are extremely frightened by media reports about the breast cancer risks of estrogen treatment.

When higher doses are employed, blood estrogen levels must be monitored and kept within a normal range as in Table 12 in chapter 8 (page 215), which lists the reference values for blood hormone levels. Again, there is no indication that breast cancer incidence differs with a dosage that ranges between 0.625 mg and 1.25 mg of conjugated estrogen, which I generally recommend for the treatment of mood, memory loss, and cognition.

THYROID HORMONE RISKS

The use of thyroid hormone to treat hypothyroidism in menopausal women entails both benefits and risks. Women should go on thyroid hormone only if blood tests confirm hypothyroidism (an underactive thyroid gland). This condition decreases the metabolic rate of all organs, which can have negative effects on the heart, brain, and hormone-secreting glands, including the ovaries. Undiagnosed hypothyroidism can even lead to a misdiagnosis of premature menopause. Women who want to raise their metabolism in order to lose weight sometimes abuse thyroid hormone. Taking too much thyroid hormone has been shown to speed the development of osteoporosis in menopausal women.

TESTOSTERONE RISKS

When I discuss adding low-dose testosterone to the hormone regimen with menopausal women who've lost their sex drive and/or cannot achieve orgasm, they usually raise a number of concerns. The first is, "Will I grow a beard?" I assure them that they've had low levels of testosterone in their blood all their adult life, and if they haven't grown a beard yet, they have little to worry about. If there is a history of increased facial hair, I alert them to watch for any sudden reappearance or worsening. The key here, as I emphasized in chapter 4, is that the low-dose testosterone that I and other clinicians recommend will not cause excessive hair growth or other signs of "virilization," such as a deepening of the voice.

Another concern is acne. Low-dose testosterone therapy rarely causes acne in a woman who has no previous history. Caution must be used if there is such a history, because even low doses can aggravate ongoing acne. Excessively high doses of testosterone can sometimes make women irritable. I usually start menopausal women on doses of

0.3 to 0.5 mg of methyltestosterone. Not infrequently, the patient and I decide to increase her dose to 0.8 mg, and in rare cases, I have gone as high as 2.5 mg. The beneficial effects of low-dose testosterone on heart disease have been discussed at length in chapter 4.

When methyltestosterone was first employed in high doses in men, there were some liver abnormalities observed. These resulted primarily from an overload on the liver. Since methyltestosterone is metabolized and the methyl group removed in the liver, large doses of methyltestosterone swamped some of the liver enzyme systems and affected the metabolism of bile. Infrequently, men on high doses of MT became jaundiced. There was no real injury or toxicity to the liver by MT and when the hormone was discontinued, the jaundice resolved completely. The doses of MT used in women are much lower than those used in men and I have never observed any liver problems in the women I have treated. However, it is advisable to periodically give follow-up blood tests for liver function.

One advantage of methyltestosterone over testosterone itself is that it is not easily converted to estrogen, a process known as aromatization. Testosterone is readily aromatized and can significantly increase serum estrogen levels in women with a history of breast cancer. This can be a disadvantage. Therefore, in women with breast cancer who are having problems with decreased libido or sexual dysfunction, methyltestosterone is by far the hormone of choice.

PROGESTERONE FOILS CANCER, BUT HAVE WE OVERDONE IT?

A woman's lifetime risk of endometrial cancer is quite small, but it rises dramatically when she receives unopposed estrogen replacement therapy—treatment without some form of progesterone. In the 1970s, a number of studies reported a sevenfold increase in the risk of endometrial cancer with ERT. However, studies in the 1980s suggest that the conversion of a normal endometrium to a malignant state is related to two factors—the dosage and duration of unopposed ERT. Adding a progesterone-like drug (progestin) to ERT has been shown to lower the risk of endometrial cancer to a level that is no different, or perhaps even less, than that of women who have never taken estrogen.

This cancer-preventing effect of progesterone revived the use of estrogen for menopausal women. As a result, the use of progesterone has gained immensely, but this has not been a problem-free development. As I argued in the last chapter, synthetic progestins, which have anti-estrogen properties, can counter some of estrogen's positive effects on heart, mood, and memory.

My colleagues and I have published two papers showing that progestins can negate the positive effects of estrogen on mood and cognitive ability in some women. The use of cyclic progestins in women receiving continuous estrogen often precipitates PMS-like symptoms. We are constantly searching to find better progestin compounds with fewer side effects. In addition, we are experimenting with different techniques of progestin administration, trying to reduce negative side effects.

As you consider a program of hormone modulation and discuss it with your physicians, make certain that you educate yourself about the forms and schedules of progesterone. Use the previous chapter as a guide, and make sure that you get a progesterone regimen that is safe, effective, and has the fewest side effects—especially when it comes to emotional, cognitive, and sexual concerns.

RALOXIFENE: ARE *SERMS* THE WAVE OF THE FUTURE?

Raloxifene, a "selective estrogen receptor modulator" (more commonly, a SERM), is being prescribed to treat women concerned about the breast and uterine cancer risks of HRT. A number of other SERMs are currently being tested and will shortly be available. This group of compounds is an offspring of the antiestrogen tamoxifen. Tamoxifen is widely used to treat women after breast cancer has been diagnosed. Recently a study has been completed showing that tamoxifen reduces the incidence of breast cancer in menopausal women. Moreover, in June 1999, a paper published in *The Journal of the American Medical Association* showed that raloxifene reduced breast cancer in menopausal women by 76 percent. No doubt, this is an encouraging finding.

I have not personally been involved in any research on SERMs, but a former colleague, Dr. V. Craig Jordan, did much of the seminal work on tamoxifen. He conducted his research on tamoxifen in animals at the Worcester Foundation for Experimental Biology in Massachusetts. At the time, I was director of the program that supported Dr. Jordan's work. SERMs such as tamoxifen have an important role in the treatment and prevention of breast cancer, but many questions arise as to their general utility for women who may not be at high risk for breast cancer, or who have not developed the disease.

Raloxifene is of particular interest because it may have fewer risks and side effects than tamoxifen. In addition to an estrogen-blocking effect in the breast, raloxifene exerts a similar effect on the lining of the uterus (endometrium), and does not increase the risk of endometrial cancer. (That is one of the worries about tamoxifen.) The elimination of concerns regarding breast cancer and uterine cancer is the major benefit of emerging SERM drugs like raloxifene. Raloxifene also has been demonstrated to have positive effects on bone density. A combination of raloxifene and alendronate (Fosamax) produces a greater increase in bone density than either raloxifene or alendronate alone. In this way raloxifene is similar to estrogen.

Unfortunately, SERM drugs also raise worrisome concerns. The current SERMs do not have the same protective effects as estrogen on the heart, and they may not prevent Alzheimer's dementia. Raloxifene does reduce cholesterol, but not as effectively as estrogen. While estrogen raises good cholesterol (HDL) by about 10 percent, raloxifene has very little stimulatory effect on HDL. Raloxiphene lowers bad cholesterol (LDL), but again, not as significantly as estrogen. Raloxifene has one advantage over estrogen in that it suppresses triglycerides, the fatty acids that stimulate the liver to produce cholesterol, while estrogen can raise triglycerides.

We don't yet have data on whether raloxifene has the same protective effect on heart disease as estrogen. However, animal studies reveal that while estrogen lowers atherosclerosis (hardening of the arteries) in monkeys by 70 percent, raloxifene administration produced no reduction and was no better than a placebo. Some data now suggest that women taking tamoxifen to prevent a recurrence of breast cancer have a lower risk of heart disease. At the moment, we just don't know whether raloxifene will have a similar protective effect.

Another question is whether raloxifene has any effects on the brain that are similar to estrogen, and whether it will lower the incidence of Alzheimer's disease, as does estrogen. Raloxifene does stimulate the growth of neural tissue, but cognitive testing comparing this SERM to a placebo in tests of cognitive function have shown no improvements.

Because raloxifene has antiestrogen properties, it does not relieve menopausal women's hot flashes. This is a real drawback, because many women start HRT in order to control hot flashes. On the other hand, raloxifene produces no breast tenderness. Breast tenderness (mastalgia) bothers one third of women taking estrogen, particularly those in their sixties and seventies and is one of the reasons older women discontinue estrogen treatment. Raloxifene does have a stimulatory effect on bone density, but the extent of the effect is considerably less than that of estrogen.

Raloxifene has only been available for clinical use for four years. The data that are generated over the next five to six years will supply useful information as to whether this SERM represents a major breakthrough, or is yet another antiestrogen with a host of drawbacks lined up against its benefits.

The whole concept behind the development of SERMs is to find an agent that mimics estrogen's stimulatory effect on certain tissues and organs (the brain, cardiovascular system, bones, etc.) and has no effect, or a blocking effect, on other tissues and organs (the breasts and uterus, colon, etc.). The result would be an agent that reduces hot flashes, mood disturbances, cognitive impairments, sexual problems, heart disease risk, and bone loss, while having no effect or even reducing the risk of cancers of the breast and uterus. Until we have the perfectly designed synthetic SERM, we will have to rely on the natural and synthetic hormones that most closely "imitate" those produced by our own bodies.

HORMONES FOR WOMEN WITH A HISTORY OF BREAST CANCER: TOO RISKY?

Currently, there is controversy as to whether women who have had breast cancer should ever use hormone replacement therapy. In the past, breast cancer was considered an absolute contraindication for HRT, although this was based on theoretical beliefs, with little hard data to support it. As the beneficial effects of estrogen on heart disease, osteoporosis and Alzheimer's disease have become more apparent, and evidence has shown that even women with breast cancer who take estrogen live longer, the question has arisen as to whether estrogen should always be withheld from breast cancer patients.

This has also become a quality-of-life issue. Some breast cancer patients who are seriously estrogen-deficient become depressed, anxious, unable to sleep, and plagued by memory loss and inability to concentrate. Having breast cancer is stressful enough; the overlay of emotional distress caused by such symptoms can make matters much worse. Frequently, these are women at the peak of their careers, poised to ascend even farther, who are seriously hampered by a hormone deficiency that would be easily corrected, if only the shadow of recurrent breast cancer was not hanging over them.

It will surprise many women to learn that the available data on recurrence of breast cancer in women using estrogen are encouraging. Women with a history of breast cancer who are receiving HRT do not appear to have an increased recurrence rate when compared to similar patients who don't use hormones. However, we need more large-scale studies to confirm this observation. In the interim, a woman must make this decision based on the available evidence, with careful consideration of risks and benefits, and in consultation with her doctors, including her cancer specialist. One factor to consider: whether the tumor is deemed "estrogen-dependent." A positive estrogen receptor test implies that the cancer is estrogen-dependent. Some physicians will not recommend hormones for breast cancer patients under any circumstances. I certainly believe that breast cancer patients should proceed with caution, but I also believe that in some cases it is a medically and ethically sound decision to commence with hormone treatment.

A survey by questionnaire of 224 breast cancer patients inquired as to the type of cancer treatment the patient had received and the presence of any symptoms of estrogen deficiency. The majority of the women were postmenopausal, and 44 percent of them were willing to consider ERT under medical supervision. Patients with more aggressive disease requiring radiation or chemotherapy were less willing to risk ERT. Among patients that had undergone surgery alone, a majority—71 percent—would consider ERT. It is not surprising that women highly bothered by typical menopausal symptoms were more willing to take estrogen. As the symptoms get worse, these women became more interested in treatment. I see such women frequently.

JANICE'S STORY: MAKING THE ESTROGEN DECISION

Janice, fifty-three, was one such patient. Eleven years earlier, she had a mammogram revealing a small suspicious area in her left breast. A biopsy was obtained, and Janis was told that it was negative for malignancy. Luckily, several weeks later when the head of the pathology department was preparing for a lecture and looking for microscopic slides of normal breast tissue, he came upon Janice's slide and was taken aback. What had been called normal breast tissue was actually carcinoma *in situ* of the breast, a form of breast cancer, which fortunately had not spread from the site of its original origin. After consulting with her surgeon, Janice decided to have a mastectomy and breast reconstruction.

At age forty-seven, Janice had a hysterectomy for bleeding fibroids, but her ovaries had not been removed. Despite this, after the surgery Janice noticed signs of estrogen deficiency. (Hysterectomy can partially compromise the blood supply to the ovaries.) In the years that followed, Janice's quality of life deteriorated drastically. She slept only a few hours each night, her energy was utterly depleted, and for the first time in her life, she lapsed into depression. Her memory faltered and her ability to concentrate was virtually nil. Janice was a children's book author, so you can imagine how devastating this was—she simply could no longer write. Though she was happily married, and had had a wonderful sex life with her husband, her interest in sex declined to a point where the joy was drained out of the experience. Vaginal dryness also made intercourse painful. Janice's story starkly illustrates how estrogen deficiency can have truly life-negating effects for some women.

Janice recognized that estrogen loss was a key factor in her troubles, yet she'd been counseled to forgo estrogen-based treatment, given her own history of breast cancer. But her quality of life had deteriorated so badly that she began to consider hormone treatment. She arrived at my office searching for guidance.

I spoke directly with the pathologist who originally diagnosed Janice's carcinoma *in situ.* He reviewed her slides, and when we spoke, again he reassured me that her tumor had not spread, and that if anyone was a candidate for HRT in such a situation, it was Janice. After a lengthy discussion of the pros and cons, Janice began taking a standard low dose of estrogen—Premarin 0.625 mg daily. After several weeks, we added low-dose testosterone to bolster her energy, mood, and sex drive.

Janice returned for a follow-up one month later. "It's a miracle!" she enthused. "I'm not depressed and I'm writing again. My memory is back and sex is a pleasure. The vaginal pain is gone. My insomnia has lifted, and I'm even having sexual dreams, which I haven't had for years."

In an interesting aside, the breast surgeon who had operated on Janice was Dr. Susan Love. Janice's experience with Dr. Love had been extremely positive. She told me a touching story about her preparation for mastectomy, that frightening time as she waited in the operating room. Dr. Love finally arrived, entering moments before Janice was anesthetized. "Open your hand," she said, seeming to sense Janice's fear. She pressed a small crystal into her palm. "Don't tell anyone," she whispered. I thought it was a lovely gesture, one of genuine compassion.

Dr. Love has written thoughtfully and quite critically of estrogen treatment for menopause. (Primarily in her book *Hormones.*) While I may disagree with many of her positions, I have no doubt that they are principled, and clearly, she's a surgeon with a heart. Dr. Love has been criticized for misstating that breast cancer is a greater killer of menopausal women than heart disease. In fact, heart disease fells nine times as many menopausal women as breast cancer. That mistake does not negate the fact that she is a highly accomplished researcher and physician who is loved by her patients. Overall, *Dr. Susan Love's Breast Book* is an excellent guide for women with or without breast cancer.

I understand that Dr. Love's primary concern is breast cancer—she experiences the tragedy associated with this disease every day, and has made it her life's work to prevent and cure it. Likewise, I recognize that my own positions may be influenced, to some extent, by the fact that every day I see the terrible effects that estrogen loss can have on mood, memory, cognition, and sexual happiness. Like Dr. Love, I'm sure I make strenuous efforts to separate my feelings about my patients and their pressing concerns from the facts about risk versus benefit. We all have to keep balancing those scales to determine what is best, in both the short- and long-term, for the patients who entrust us with their care.

10

Taking Charge of Your Hormones

THE DECISION IS YOURS

Perhaps you've read this entire book, identified your symptoms as being caused by hormone deficits, and decided that you wish to pursue hormone modulation. How can you get the treatment that you need? Today, many primary care doctors and gynecologists are not always aware of the recent research on hormones and the mind. But there are many gynecologists who are well aware of the hormone/mind connection and who indeed are involved actively in research on the effects of hormones on the brain. Dr. Frederick Naftolin, the Chief of Obstetrics and Gynecology at Yale University and Dr. James Simon, of the Women's Health Research Center in Maryland, are two of the leading researchers in the effects of estrogen on brain function, mood, memory, and cognition. However, research takes a long time to reach the practicing clinician who is reluctant to stray too far from well-established medical paths. Such physicians may reserve HRT for women after menopause, and rely on the standard "one size fits all" combination of estrogen and progestin. Among these doctors, individualized hormone modulation—following blood levels and changing hormonal preparations and doses—is not part of their therapeutic strategy.

But in my experience, many primary care physicians, gynecologists, endocrinologists, and psychiatrists are open and flexible when made aware of scientific facts and are willing to discuss solid research-based innovative approaches to treatment. You can put together your own program of hormone modulation with this approach: Arm yourself with information, communicate with your doctors, help them to communicate with one another, and assert your needs. If your health care providers do not listen or respond, or they're not sufficiently flexible, find someone who will listen and is not dogmatic.

I don't mean to claim that this process will be easy, but once you are committed, you should be able to get the hormone modulation you need for optimal health and well-being.

Women are advised to weigh their various risk factors—breast and uterine cancer, heart disease, osteoporosis, etc.—when considering HRT for menopausal symptoms. The same risk/benefit calculations should apply when women pursue hormone treatment for symptoms of the mind. Still, many doctors hesitate to consider hormones for mental well-being due to concerns about breast cancer risk, though they have far less hesitation when it comes to HRT for other menopausal symptoms. The reason? Many doctors are simply not aware of the vast potential of hormone therapy for psychological health. It is also possible that doctors trained to be sensitive to physical complaints such as hot flashes and diseases such as osteoporosis are not as sensitive to the suffering associated with mood, memory, and sexual disorder. They often do not wish to become involved in the care of patients whom they identify as having a mental or emotional condition.

Sadly, the prejudice against mental and emotional conditions is present in the lay population as well as in some portion of the medical community. I am frequently told by my patients, "No one understands me. My husband and my children can't understand why I am acting this way. I feel like I am going crazy." Wives often ask if they can bring their husbands in so I can explain to them their wives' menopausal symptoms. I take great pains to assure them that they are not going crazy, that their mood and emotional complaints have a physical basis correctable with hormonal treatment. I tell them, "It is not my job to change you, but simply to allow you to once again be yourself."

BRINGING TOGETHER SPECIALISTS

For those who seek a whole-person approach to medicine, the traditional boundaries of psychology, psychiatry, neurology, and endocrinology must be dismantled. It is now well established that hormones affect the bones, the heart, the eyes, and the brain. Stress, genetic factors, or environmental influences can cause hormonal fluctuations, which in turn affect the brain. Even the immune system—our body's inner healer—can be altered by changes in hormones and brain chemistry. This means that specialists must combine their areas of expertise to come up with the best understanding and treatment of diseases with multiple causes.

However, for the patient, finding treatment for multiple-level disorders can be a challenge. There are very few clinical psychoneuroendocrinologists practicing anywhere in the country, and you won't find a listing in your local phone book. As I've emphasized, you have to take charge of your own health care in order to assemble a medical team that can provide individualized hormone modulation for your conditions.

Many adults have only one doctor, usually a family physician, internist, or obstetrician/gynecologist, who may be too busy or feel inadequately trained to respond to the patient's cognitive, mood, or memory complaints. Managed care doesn't help, putting pressure on doctors to cut down on time with each patient. Patients are often interrupted before they can even tell their doctors about all of their symptoms and concerns.

A 1999 report in the *Journal of the American Medical Association,* a survey of 264 patient-physician interviews, revealed that only 28 percent of physicians let their patients inform them of the full spectrum of their health complaints.[1] Instead, the patient is cut short and the doctor focuses on just one concern. The authors of this article determined that if the patients were given approximately six seconds more, most would supply the physician with all the information needed for the most efficient and accurate diagnosis and treatment. I would suggest that patients need more time than that, but the study reveals just how truncated the patient/doctor relationship has become.

I recently had a patient, Elena, who was suffering with depression that began immediately prior to her menopause. Elena had never been depressed before, though she did have a family history of depression. She brought her husband to my office so I could help him understand what was happening to her. I spent considerable time explaining the link between hormones and depression, answering all their questions. Just before the couple left, Elena turned to me and said, "You're the only doctor I feel I have had a conversation with." I replied, "I can't tell you how sorry I am to hear that."

By way of explanation, I told her that it's part of my job as an internist to effectively communicate. Not all physicians are internists.

Many physicians have functions to fulfill that are not required of an internist. Surgeons, who spend much of their time in the operating room, usually interact in a different way with their patients. I don't excuse poor communication skills, but in our current era of specialization and managed care, they're understandable. I explain to my patients that highly skilled surgeons save lives but cannot spend a great deal of time talking to their patients. But these "hands-on" physicians perform a vital job. As I said to Elena and her husband, "Please don't ask me to operate on you; you'd be in big trouble."

The only way to achieve your needs is to find open-minded physicians and specialists and then to use your own communication skills to insist that they consider your specific needs and concerns when it comes to hormones for the mind.

IDENTIFY YOUR MEDICAL FACILITATOR

All you may need is one physician or specialist to listen, communicate, and recognize the validity of your concerns in order to initiate the dialogue among specialists that will yield proper treatment. This individual is your medical facilitator—the doctor who coordinates your individualized hormone modulation. It can be your primary care physician, internist, or surgeon who holds your best interests at heart. It could also be a psychiatrist, gynecologist, or endocrinologist, but I've often found the primary care doctor most adept at jump-starting the necessary cross-disciplinary communication.

For instance, I recently had a patient who was about to undergo minor surgery. She was taking an MAO inhibitor as an antidepressant, which can have fatal consequences when mixed with certain forms of surgical anesthesia. Because she could not get her psychiatrist, surgeon, and anesthesiologist to talk with one another, the surgery was delayed three months. It was not until her primary care doctor intervened that the necessary dialogue ensued and the problem was solved, enabling her to undergo surgery.

I encourage you to identify the doctor you trust to help you develop your own regimen of individualized hormone modulation. You want a physician willing to listen and learn, particularly if he or she is not aware that hormones have an influence on mood. If you have a trusted primary care physician, see first whether you can enlist him or her in your effort to procure proper hormone modulation. If he or she is not the ideal facilitator, consider your gynecologist. If no one on your current medical team is sufficiently responsive, you must begin searching for doctors who can help.

Finding the appropriate doctor(s) will require some research. The Board of Registration of Medicine in some states has physician profiles accessible through the Internet. You can learn about a physician's training, area of specialty, malpractice record, and scientific publications. Go to the Internet for help, including the American Medical Association's Web site (www.ama-assn.com) and use their doctor finder service, which enables you to plug in specialties and geographic locations to identify the right specialist.

I urge patients to keep searching, if necessary, to find a doctor to orchestrate their care. It is also important that the physician be aware of the most recent advances in hormone treatment. He or she must be willing to adequately adjust hormone doses to treat mood, cognitive, and sexual disorders. The critical part of this process requires the measurement of blood hormone levels, and patients should know that many physicians are not accustomed to obtaining such measurements. If there is a real rigidity and an unwillingness to listen and learn on the part of the physician, you may need to look elsewhere.

EDUCATE YOUR PHYSICIANS

You should feel okay telling your doctor that you've conducted your own medical research. Most doctors are happy to have well-educated patients who can intelligently discuss their medical care. They recognize that in the Internet age, patients are frequently on top of recent developments. You may wish to use this book as a resource and guide for your doctor, pointing to the research findings amassed here to frame your discussions.

Doctors are usually willing to consider different clinical approaches when there is sound published research to back them up. The numerous studies on the value of hormone modulation for mood, memory, and sexual function—all of them cited in the pages of this book—should enable the members of your medical team to feel comfortable initiating and guiding your treatment. Moreover, there is ample research supporting the safety and efficacy of varying dosages and types of estrogen, progesterone, and testosterone—the basis of flexible hormone modulation.

Consider one of the most common problems women face in this regard. Peri- and postmenopausal women with a stubborn depression may have a hard time getting proper treatment, such as a combination of an antidepressant and hormone modulation (see chapter 2). Why the difficulty? The family physician or gynecologist often refers a depressed patient to a psychiatrist. Most psychiatrists have limited training in endocrinology, and thus are unlikely to consider a hormonal link to the patient's depressed mood. An endocrinologist understands hormones but is not necessarily trained to diagnose depression. The endocrinologist may not even recognize the far-reaching effects of hormones on the mind.

COMMUNICATE ASSERTIVELY

The patient will need to establish assertive communication with doctors and specialists who will make up the medical "team" they need to get appropriate hormone treatment. This may require only two doctors—the primary care physician and one other specialist—or as many as three or four. Again, it will help if one individual acts as the medical facilitator.

The secret to successful assertive communication is to be both proactive and respectful. Don't avoid straightforward descriptions of your symptoms, and don't shy away from letting the doctor(s) know how much you know. Be completely open about your suspicions. For instance, if you're depressed and your antidepressant isn't working, be willing to suggest estrogen deficiency as an explanation, if that is what you believe. If you are forty-five and having memory lapses, be willing to suggest perimenopause as a factor. Don't hesitate to share your thoughts and theories, as well as your symptoms.

Be respectful. Because patients often feel intimidated by physicians, they forget that doctors are no different from anyone else in this regard: They wish to be treated with decency and dignity. So be straightforward and assertive, but if you wish to build a productive partnership, don't hurl accusations of insensitivity.

Avoid making demands about the medications you think might be right for you. Some physicians are understandably sensitive about being pressured into writing prescriptions they may or may not believe are appropriate. Build a relationship based on mutual respect and trust, and you will enable yourself to get the treatment you need.

Write down your symptoms, concerns, thoughts, and questions before speaking with members of your medical team. Keep them brief and to the point, recognizing realistically that you may have only a few minutes of conversation with a particular doctor. Know exactly what you want from an interaction, and prepare ahead of time so that you can achieve your goals.

It is common practice for most physicians to communicate frequently with their peers in various specialties, but communications across certain specialties are less common. Communication between a gynecologist and a neurologist may be quite effortless but communication between a gynecologist and a psychiatrist or a psychologist may require more effort on the part of both doctors. They don't share a common medical language and some physicians may even have a bias against dealing with what they identify as mental illness. This bias may even be seen in medical journals where researchers are communicating scientific information to other physicians.

THE IMPORTANCE OF MEDICAL CROSS-TALK: MY EXPERIENCE

In the June 1999 issue of the *Journal of Clinical Endocrinology and Metabolism* there were numerous articles discussing HRT. The experts writing this report made very careful and well-thought-out commentaries on the risks and benefits of estrogen for the treatment of heart disease and osteoporosis. It was as if the heart and bones existed alone from the rest of the body with no connection to the brain. A few remarks were made regarding HRT and breast cancer. How the brain's response to estrogen deficiency can affect a woman's quality of life was ignored. Estrogen's beneficial action on the heart was estimated in one of these articles to extend life for over three years. Effects on the brain, which is extremely sensitive to estrogen deficiency, could well affect a woman's quality of life for over thirty years. The blinders of specialists are as evident in scientific research as in clinical practice.

My experience in the past thirty years has lead me to believe that communications between an endocrinologist and a psychiatrist can be extremely fruitful, beneficial to the patient, and educational for the doctors. My own personal experience began with my work with Don Broverman and Bill Vogel, both Ph.D.s. Our conversations grew into a lifelong collaboration. This has so positively colored my feelings about medical cross-talk that I may be overly optimistic about the success of such a process. Nevertheless, I feel that specialists from differing fields should actively pursue contact.

I converse with a psychiatrist, Dr. Mary Collins at McLean Hospital in Belmont, Massachusetts, on an almost weekly basis (sometimes two or three times a week). Dr. Collins has taught me a great deal of what I know about modern psychiatry, as we have collaborated on the care of numerous patients. Indeed, she has diagnosed many of my patients with psychiatric disorders, which has made it possible for them to get the care they needed. Dr. Collins is also knowledgeable in the areas of endocrinology and neurology, and is a skilled communicator who has no difficulty speaking with physicians in these specialties.

Recently Dr. Collins and I shared responsibility for the care of a woman whose manic depression worsened as she entered menopause. She was having hot flashes and night sweats and clearly needed estrogen treatment. My previous experience had warned me, however, of the possible dangers of simply giving hormones to this patient. Estrogen treatment for manic-depressive patients can exacerbate their symptoms, even leading to a manic psychosis. I needed the advice and collaboration of an experienced psychopharmacologist. When I called Dr. Collins, she agreed that treating this patient had potential pitfalls. She believed that if she instituted treatment with a mood stabilizer such as lithium or Depakote, that estrogen would probably be safe. That is exactly what transpired. My patient saw Dr. Collins, who prescribed Depakote. Only then did I start the patient on estrogen therapy, without incident. In fact, the woman's cyclical mood disturbance subsided, and without her menopausal symptoms, she was able to reclaim a sense of well-being.

In some cases, communication between doctors can grow into a fruitful collaboration. About ten years ago, I exchanged views on the use of testosterone for the treatment of reduced sexual libido and function in menopausal women with the Boston psychiatrist Dr. Susan Rako. (See chapter 4.) Dr. Rako was writing a book on this subject (*The Hormone of Desire*) and she wished to discuss my experience administering testosterone to women and men. Her own personal experience during menopause had led her to believe that testosterone was by far the most effective treatment for the sexual problems of menopause. However, we both felt that the doses of testosterone commercially available for women were too high. Dr. Rako had solved this problem by having testosterone compounded by a compounding pharmacist, Steven Grossman at the Pierce Pharmacy in Brookline, Massachusetts. Before that, I had instructed my patients to cut testosterone tablets into halves and quarters, but I began using the compounded methyltestosterone. Since that time, Dr. Rako and I have shared many patients in one of the most rewarding collaborations in my professional experience.

The productive outcome of my collaborations with psychiatrists, psychologists, and gynecologists is an example of the kind of communication that must take place between specialists in differing fields. As we pass into the new millennium, we need just this type of communication to successfully care for our patients, whose conditions cannot be simply pigeonholed into one narrow category.

A patient—in this case, the woman with hormone-associated problems of mood, memory, cognition, or sexuality—can ultimately help bring about the medical cross-talk that will break down the old barriers between specialties. She should do so for her own well-being, to get the medical care she needs for her multileveled medical conditions. The long-range result will be a much more responsive and effective medicine, one that diagnoses and treats the whole woman.

11

New Medicine for the Mind

THE FUTURE OF HORMONE MODULATION

For many women, hormonal treatment should begin as soon as possible after the emergence of early symptoms of hormonal imbalance. This is especially crucial during the transition from perimenopause to menopause. As women advance toward menopause, their estrogen levels decline much more dramatically. Since estrogen stimulates its own receptors on brain and other cells, a woman who has been menopausal for six years will not be as responsive to hormone modulation as a woman who has been menopausal for only two years. Fortunately for patients, a number of exciting developments are helping to bring hormone modulation into the mainstream mental health treatment.

- In 1979, my colleagues and I published our finding of the antidepressant effect of estrogen in severely depressed, treatment-resistant women.[1] While this publication was widely cited, the use of estrogen as an antidepressant never became popular. The main reason was concern in the mid-1970s about estrogen and uterine cancer. The use of progesterone has changed that risk so that women on estrogen are at no more risk of uterine cancer than women receiving no treatment. As a result, interest in estrogen's many benefits was revived. Recent studies have shown that menopausal women who were not on hormone replacement did not respond well to the antidepressant actions of Prozac, while women on HRT had a high success rate—about 75 to 80 percent of the women experienced significant relief of their depressive symptoms.[2]

- Depressed middle-aged men who had low testosterone levels showed little response to antidepressants similar to Prozac. When testosterone was then added to the antidepressant regimen the vast majority of men experienced a dramatic recovery from major depression. Substitution of a placebo for the testosterone resulted in recurrence of the depression in 75 percent of these men.

Such encouraging results have caused a stir among many leaders in psychiatry, endocrinology, and neurology. A renowned psychopharmacologist, Dr. Stephen M. Stahl at the Clinical Neuroscience Research Center in San Diego, California, has recently referred to the profound effects of estrogen on behavior. He has called upon the modern psychiatrist evaluating women to obtain a complete reproductive history, including details of hormone treatments, while identifying reproductive events as triggers of various psychiatric disorders. While I certainly think that more large-scale studies are needed to help guide the use of estrogen with antidepressants, I believe that current information about estrogen's positive effects on the nervous system safely allows physicians treating depressed menopausal women to include hormone modulation as part of their armamentarium. Other areas of research have stimulated the growing interest in estrogen's effects on the brain:

- *Neurotransmitter/hormone relationships:* Hundreds of studies point to the interactions between hormones, especially estrogen, and the key neurotransmitter systems in the brain responsible for balanced moods, sharp cognition, sound memory, and sexual function. Our understanding grows with each month and the release of more basic and applied research. Soon we will learn more about the full extent of the interconnections of our endocrine and nervous systems. When we do, the use of hormone treatments, alone and in combination with psychiatric and other medications, will become both more precise and widespread.
- *Chemical, electrical, and circulatory activity in the brain:* New research on the influence of hormonal factors on neurotransmission, cerebral blood flow, and electrical activity in the brain is heralding a revolution in our understanding of how biological systems interact to influ-

ence the "mind," with all this entails in terms of mood, cognition, and sexual health.

- *Alzheimer's disease:* The use of relatively high doses of estrogen and prolonged treatment regimens has resulted in a reduction of Alzheimer's disease by 80 percent.[3] This is an astounding figure in the case of a disease that, until now, has resisted so many treatment approaches. Furthermore, when doctors combine estrogen with certain anti-Alzheimer's drugs, such as tacrine or Aricept, patients have experienced marked improvements in their memory and cognitive abilities.[4]
- *Parkinson's disease:* Estrogen treatment also has been reported to decrease the incidence of Parkinson's disease and prevent the development of dementia.[5] The protective effects of estrogen on Alzheimer's and Parkinson's disease have generated intense interest in brain-hormone research. Studies from the department of neurology at Beth Israel Medical Center in New York City report that estrogen use is associated with lower symptom severity in women with early Parkinson's disease. The National Institutes of Health and pharmaceutical companies are beginning to generously fund these research avenues.

Reports on hormones and the brain are appearing in an ever-widening swath of prestigious peer-reviewed medical journals. Medical textbooks on the effect of hormones on the brain are being published and hold the promise that hormone therapy for mood, memory, and cognitive diseases may become a required subject in medical schools, and a cross-disciplinary practice that will soon be applied in clinics around the country.

A recent fact sheet issued by the Research Initiatives Committee and the Public Communications Committee of the Endocrine Society was published on the Endocrine Society's Web site (www.endo-society.org). It addressed the relationship between endocrinology and menopause. The point was made that endocrine research plays an important role in addressing the many questions that need to be answered about natural and surgical menopause. The following six questions were listed:

- When women with functioning ovaries have to have their uterus removed, should the ovaries be removed? If so, should the age of the women affect their decision?

- How do the sex steroids relate to cancer and the functioning of the immune system and the brain?
- In what combination, amounts, and route should hormone replacement therapy be given? How should we determine and monitor the doses that a particular woman needs?
- Will the use of estrogen and progesterone in menopause increase women's productivity, functioning, and quality of life and decrease health problems and death as predicted? What is the cost of this treatment?
- How do other health problems like breast cancer affect the risks and benefits of hormone replacement in a woman?
- What are the best ways to monitor ovarian function?

When I read these questions, I was pleased to realize that this book had addressed all of them in detail. While further research is certainly needed in all of these areas, I have attempted to provide you with the most current available information, and to interpret and evaluate its usefulness so that you will be able to benefit from the hormone revolution that awaits us in the not-too-distant future.

DOCTOR-DOCTOR AND DOCTOR-PATIENT COMMUNICATION: THE KEY TO CHANGE

One of our greatest challenges for the future development of hormone modulation is to improve communication among doctors in different fields. One recent case brought this home to me in very practical terms. I had spoken with one of my patients, Maria, an older woman who had, in the past two years, lost her husband, a son, and a brother. She'd been remarkably resilient in coping with these losses, but in recent months, the full impact of their deaths finally hit her full force, resulting in a clinical depression. She had difficulty sleeping, and her psychiatrist prescribed a small dose of a tricyclic antidepressant. This treatment worked reasonably well until she began to develop an irregularity of her pulse. Maria's cardiologist told her to stop taking the antidepressant, which sometimes can aggravate an arrhythmia. He gave her a thorough cardiac evaluation, but found no serious heart defect.

Predictably, Maria's depression and insomnia came right back. Thus began a merry-go-round of antidepressant trials. Unfortunately, none did the job effectively without intolerable side effects. She would regularly wake at 4 A.M., hounded by painful memories. It finally became clear that the original tricyclic antidepressant was Maria's best option. But she wasn't sure how to go about reinstating this form of treatment, given the supposed risks. Would it really be dangerous? she wondered. The psychiatrist assumed that the cardiologist would oppose the resumption of this medication. The cardiologist had no idea of Maria's severe difficulties finding a satisfactory antidepressant. Clearly, communication between the psychiatrist and cardiologist—or at least a liaison between the two—was necessary.

When a bewildered Maria finally consulted me, a simple phone call to the psychiatrist and cardiologist helped to resolve the problem. The cardiologist reviewed her records and felt that her arrhythmia was of minimal medical significance. In view of her psychiatric problems, he was not opposed to her resuming her previous treatment. Once this information was conveyed to the psychiatrist, a happy solution to the dilemma was achieved. We need a new cadre of physicians who are generalists, capable of just this kind of cross-disciplinary communication. These specialists must not only learn about other overlapping areas of medicine (i.e., the psychiatrist and endocrinologist should know more about each other's fields), they must be educated in communication skills that enable them to exchange information and ideas with each other and with their own patients.

While psychiatrists, psychologists, neurologists, and endocrinologists often stay within the narrow confines of the biological system they understand, the patient's body does not recognize such borders. We can no longer be considered a collection of discreet organ systems; our mental state is indivisible from the functions of our nervous, endocrine, and immune systems. Specialists must dismantle artificial walls between their worlds and learn to speak the same language for their patients' sake. It also is the responsibility of medical schools and medical educators to train new physicians to be better communicators.

For instance, psychiatrists at McLean Hospital near Boston, where psychiatrists affiliated with Harvard Medical School hospitalize their patients, often refer patients to me. Several times a week, I conduct phone consultations with the psychiatrists to coordinate the hormonal and psychopharmacological treatment of these patients. This sort of dialogue is crucial in a field of medicine that traverses several medical specialties.

Ultimately, I believe patients who take their health into their own hands will be the ones to educate and inspire their doctors to become more open and communicative, and the patients themselves will ultimately benefit the most. Individualized hormone modulation is both a model for whole-person medicine and a solution to vexing problems of mood, memory, and sexuality. Millions of women and countless men will soon be the beneficiaries.

YOUR HORMONE DECISION: ADD BRAIN HEALTH TO YOUR CALCULUS

The most important message in this book is simple: Add brain health to your calculus as you make decisions about hormones. Women are rightly counseled by their doctors and by media medical experts to consider these primary concerns: heart health, bone health, menopausal symptom relief (i.e., hot flashes, night sweats, vaginal dryness, etc.), and cancer risk. But we have learned—and continue to learn—that women's hormonal profile during perimenopause and menopause has a major impact on their mood states, cognitive functions, memory, and sexual well-being.

Why should these aspects of women's health be any less important than the other factors? They should not be. While some hormonal considerations may affect length of life—namely, heart and cancer risks—others affect quality of life, including such symptoms as hot flashes. For women with hormone-related disorders of mood, cognition, and sexuality, the impact on quality of life is arguably more serious than that of the physical symptoms of menopause. For women, the quality of the second half of their lives may depend on their brain health, so I see no reason not to rank this factor right at the top of the list with heart disease, cancer, osteoporosis, and the other quality-of-life variables. Moreover, one brain disease—Alzheimer's—is indeed life-threatening, and we are continuing to learn how hormonal balance plays a crucial role in reducing the risk of this devastating disease.

There is no conspiracy to ignore the relationship between hormones and the brain and the clinical applications that apply. But I do think doctors and patients must educate themselves and keep up to date on recent developments that underscore the importance of hormones in brain health. The issues related to hormones and the mind have taken longer to reach the consciousness of the medical community and the public, in large part because researchers have had to overcome certain false assumptions: that hormones do not cross the blood-brain barrier, that they do not influence neurotransmission, that they cause terrible increases in cancer risk, that they have no place in the treatment of psychiatric, cognitive, and sexual disorders. With new research and clinical findings, these false assumptions are dropping away, one at a time.

Now, we need to embrace a brand-new assumption, one that wipes away the old, wrongheaded notions about hormones and the mind. I can sum it up in one sentence: *Hormones matter to brain health, and brain health matters to women.* If this book makes any contribution to the way women and their doctors practice health care, I hope it will be the acceptance of this basic truth.

References

Introduction

1. Klaiber, E. L., D. M. Broverman, et al. (1967). "The automatization cognitive style, androgens, and monoamine oxidase." *Psychopharmacologia* 11(4): 320–36; Klaiber, L. E., M. D. Broverman, et al. (1974). Rhythms In Cognitive Functioning and EEG Indices In Males. *Biorhythms and Human Reproduction.* M. Ferin. New York, N.Y., John Wiley & Sons: 481–93.

2. Klaiber, E. L., D. M. Broverman, et al. (1974). "Rhythms in plasma MAO activity, EEG, and behavior during the menstrual cycle." *Biorhythms and Human Reproduction.* M. Ferin. New York, N.Y., John Wiley & Sons: 353–67; Klaiber, E. L., D. M. Broverman, et al. (1979). "Estrogen therapy for severe persistent depressions in women." *Arch Gen Psychiatry* 36(5): 550–54.

3. Grodstein, F., M. J. Stampfer, et al. (1997). "Postmenopausal hormone therapy and mortality [see comments]." *N Engl J Med* 336(25): 1769–75.

4. Nachtigall, L. E., R. H. Nachtigall, et al. (1979). "Estrogen replacement therapy II: a prospective study in the relationship to carcinoma and cardiovascular and metabolic problems." *Obstet Gynecol* 54(1): 74–79; Nachtigall, M. J., S. W. Smilen, et al. (1992). "Incidence of breast cancer in a 22-year study of women receiving estrogen-progestin replacement therapy [see comments]." *Obstet Gynecol* 80(5): 827–30; Helzlsouer, K. J. and R. Couzi (1995). "Hormones and breast cancer." *Cancer* 76(10 Suppl): 2059–63.

5. Grodstein, Stampfer, et al.

6. Gapstur, S. M., M. Morrow, et al. (1999). "Hormone replacement therapy and risk of breast cancer with a favorable histology: results of the Iowa Women's Health Study [see comments]." *Jama* 281(22): 2091–97.

7. Hammond, C. (1998). *Therapeutic Options For Menopausal Health—Monograph.* Durham, N.C., Duke University Medical Center and MBK Associates, LLC.

8. The Writing Group for the PEPI Trial (1996). "Effects of hormone therapy on bone mineral density." *JAMA* 276: 1389–96.

9. Furner, S. E., F. G. Davis, et al. (1989). "A case-control study of large bowel cancer and hormone exposure in women." *Cancer Res* 49(17): 4936–40; Grodstein, F., M. E. Martinez, et al. (1998). "Postmenopausal hormone use and risk for colorectal cancer and adenoma [see comments]." *Ann Intern Med* 128(9): 705–12.

10. Paganini-Hill, A., R. K. Ross, et al. (1988). "Postmenopausal oestrogen treatment and stroke: a prospective study." *Bmj* 297(6647): 519–22.

11. The Eye Disease Case-Control Study Group (1992). "Risk factors for neovascular age-related macular degeneration. The Eye Disease Case-Control Study Group." *Arch Ophthalmol* 110(12): 1701–8.

12. Henderson, V. W., A. Paganini-Hill, et al. (1994). "Estrogen replacement therapy in older women. Comparisons between Alzheimer's disease cases and nondemented control subjects." *Arch Neurol* 51(9): 896–900.

13. Zweifel, J. E. and W. H. O'Brien (1997). "A meta-analysis of the effect of hormone replacement therapy upon depressed mood [published erratum appears in *Psychoneuroendocrinology* 1997 Nov;22(8):655]." *Psychoneuroendocrinology* 22(3): 189–212.

14. Sherwin, B. B. (1998). "Estrogen and cognitive functioning in women." *Proc Soc Exp Biol Med* 217(1): 17–22.

15. Sarrel, P., B. Dobay, et al. (1998). "Estrogen and estrogen-androgen replacement in postmenopausal women dissatisfied with estrogen-only therapy. Sexual behavior and neuroendocrine responses." *J Reprod Med* 43(10): 847–56.

1. Deep Impact: Hormones and the Brain

1. Fink, G., B. E. Sumner, et al. (1996). "Estrogen control of central neurotransmission: effect on mood, mental state, and memory." *Cell Mol Neurobiol* 16(3): 325–44.

2. Stahl, S. M. (1998). "Augmentation of antidepressants by estrogen." *Psychopharmacol Bull* 34(3): 319–21.

3. Lane, R. M. (1997). "A critical review of selective serotonin reuptake inhibitor-related sexual dysfunction; incidence, possible aetiology and implications for management." *J Psychopharmacol* 11(1): 72–82.

4. Stahl, S. M. (1996). "Depression." *Essential Psychopharmacology.* New York, N.Y., Cambridge University Press.

5. Henderson, V. W. (1997). "The epidemiology of estrogen replacement therapy and Alzheimer's disease." *Neurology* 48(5 Suppl 7): S27–35.

6. Saunders-Pullman, R., J. Gordon-Elliott, et al. (1999). "The effect of estrogen replacement on early Parkinson's disease." *Neurology* 52(7): 1417–21; Kompoliti, K. (1999). "Estrogen and movement disorders." *Clin Neuropharmacol* 22(6): 318–26.

7. Pfaff, D. W. (1968). "Autoradiographic localization of testosterone-3H in the female rat brain and estradiol-3H in the male rat brain." *Experientia* 24(9): 958–59; Pfaff, D. W. (1968). "Uptake of 3H-estradiol by the female rat brain. An autoradiographic study." *Endocrinology* 82(6): 1149–55; Pfaff, D. W. (1968). "Autoradiographic localization of radioactivity in rat brain after injection of tritiated sex hormones." *Science* 161(848): 1355–56.

8. Stahl, S. M. (1996). "Principles of Chemical Neurotransmission."

9. Stahl, S. M. (1996). Receptors and Enzymes as the Targets of Drug Action. *Essential Psychopharmacology.* New York, N.Y., Cambridge University Press.

10. Delgado, P. L. (2000). "Depression: the case for a monoamine deficiency." *J Clin Psychiatry* 61(Suppl 6): 7–11.

11. Schildkraut, J. (1965). "The catecholamine hypothesis of affective disorders: A review of supporting evidence." *Am J Psychiatry* 122: 508–22.

12. Cardinali, D. P. and E. Gomez (1977). "Changes in hypothalamic noradrenaline, dopamine and serotonin uptake after oestradiol administration to rats." *J Endocrinol* 73(1): 181–82; Farr, S. A., W. A. Banks, et al. (2000). "Estradiol potentiates acetylcholine and glutamate-mediated post-trial memory processing in the hippocampus." *Brain Res* 864(2): 263–69.

13. Ferrer, M., M. Meyer, et al. (1996). "Estrogen replacement increases beta-adrenoceptor-mediated relaxation of rat mesenteric arteries." *J Vasc Res* 33(2): 124–31; Nimmo, A. J., E. M. Whitaker, et al. (1989). "The presence of beta-adrenoceptors in rat endometrium is dependent on circulating oestrogen." *J Endocrinol* 122(2): R1–4.

14. Woolley, C. S. and B. S. McEwen (1993). "Roles of estradiol and progesterone in regulation of hippocampal dendritic spine density during the estrous cycle in the rat." *J Comp Neurol* 336(2): 293–306.

15. Brinton, R. D., P. Proffitt, et al. (1997). "Equilin, a principal component of the estrogen replacement therapy premarin, increases the growth of cortical neurons via an NMDA receptor-dependent mechanism." *Exp Neurol* 147(2): 211–20.

16. Resnick, S. M., P. M. Maki, et al. (1998). "Effects of estrogen replacement therapy on PET cerebral blood flow and neuropsychological performance." *Horm Behav* 34(2): 171–82; Berman, K. F., P. J. Schmidt, et al. (1997). "Modulation of cognition-specific cortical activity by gonadal steroids: a positron-emission tomography study in women." *Proc Natl Acad Sci U S A* 94(16): 8836–41.

17. Shaywitz, S. E., B. A. Shaywitz, et al. (1999). "Effect of estrogen on brain activation patterns in postmenopausal women during working memory tasks." *Jama* 281(13): 1197–202.

18. Zweifel, J. E. and W. H. O'Brien (1997). "A meta-analysis of the effect of hormone replacement therapy upon depressed mood [published erratum appears in *Psychoneuroendocrinology* 1997 Nov;22(8):655]." *Psychoneuroendocrinology* 22(3): 189–212.

19. Sherwin, B. B. (1996). "Hormones, mood, and cognitive functioning in postmenopausal women." *Obstet Gynecol* 87(2 Suppl): 20S–26S; Sherwin, B. B. and M. M. Gelfand (1985). "Sex steroids and affect in the surgical menopause: a double-blind, cross-over study." *Psychoneuroendocrinology* 10(3): 325–35.

20. Sherwin, B. B. (1994). "Estrogenic effects on memory in women." *Ann N Y Acad Sci* 743: 213–30; discussion 230–31.

21. Persky, H., L. Dreisbach, et al. (1982). "The relation of plasma androgen levels to sexual behaviors and attitudes of women." *Psychosom Med* 44(4): 305–19.

22. Sherwin, B. B. (1991). "The impact of different doses of estrogen and progestin on mood and sexual behavior in postmenopausal women." *J Clin Endocrinol Metab* 72(2): 336–43.

23. Klaiber, E. L., D. M. Broverman, et al. (1971). "Effects of infused testosterone on mental performances and serum LH." *J Clin Endocrinol Metab* 32(3): 341–49.

24. Vogel, W., E. L. Klaiber, et al. (1985). "A comparison of the antidepressant effects of a synthetic androgen (mesterolone) and amitriptyline in depressed men." *J Clin Psychiatry* 46(1): 6–8.

25. Henley, W. N. and T. J. Koehnle (1997). "Thyroid hormones and the treatment of depression: an examination of basic hormonal actions in the mature mammalian brain." *Synapse* 27(1): 36–44.

26. Martinez-Weber, C., P. F. Wallack, et al. (1993). "Prevalence of thyroid autoantibodies in ambulatory elderly women." *Mt Sinai J Med* 60(2): 156–60.

27. Weissel, M. (1999). "[Administration of thyroid hormones in therapy of psychiatric illnesses]." *Acta Med Austriaca* 26(4): 129–31.

28. Whybrow, P. C. (1994). "The therapeutic use of triiodothyronine and high dose thyroxine in psychiatric disorder." *Acta Med Austriaca* 21(2): 47–52.

2. Reclaiming Emotional Balance: Depression and Mood

1. Kessler, R. C., K. A. McGonagle, et al. (1993). "Sex and depression in the National Comorbidity Survey. I: Lifetime prevalence, chronicity and recurrence." *J Affect Disord* 29(2–3): 85–96; Kessler, R. C., K. A. McGonagle, et al. (1994). "Lifetime and 12-month prevalence of DSM-III-R psychiatric disorders in the United States. Results from the National Comorbidity Survey." *Arch Gen Psychiatry* 51(1): 8–19.

2. Romano, S., R. Judge, et al. (1999). "The role of fluoxetine in the treatment of premenstrual dysphoric disorder." *Clin Ther* 21(4): 615–33; discussion 613.

3. Pariser, S. F., H. A. Nasrallah, et al. (1997). "Postpartum mood disorders: clinical perspectives." *J Womens Health* 6(4): 421–34.

4. Ahokas, A., M. Aito, et al. (2000). "Positive treatment effect of estradiol in postpartum psychosis: a pilot study." *J Clin Psychiatry* 61(3): 166–69.

5. Bloch, M., P. J. Schmidt, et al. (2000). "Effects of gonadal steroids in women with a history of postpartum depression [In Process Citation]." *Am J Psychiatry* 157(6): 924–30.

6. Longcope, C. (1990). "Hormone dynamics at the menopause." *Ann N Y Acad Sci* 592: 21–30.

7. Schmidt, P. J., C. A. Roca, et al. (1997). "The perimenopause and affective disorders." *Semin Reprod Endocrinol* 15(1): 91–100.

8. Kaufert, P. A., P. Gilbert, et al. (1992). "The Manitoba Project: a re-examination of the link between menopause and depression [see comments]." *Maturitas* 14(2): 143–55.

9. Barbach, L. G. (1994). *The Pause: Positive Approaches to Menopause,* New York, Penguin USA.

10. Spitzer, R. L., Endicott, J., Robins, E. (1977). *Research diagnostic criteria (RDC) for a selected group of functional disorders.* Biometric Research, New York State Psychiatric Institute.

11. Zweifel, J. E. and W. H. O'Brien (1997). "A meta-analysis of the effect of hormone replacement therapy upon depressed mood [published erratum appears in *Psychoneuroendocrinology* 1997 Nov;22(8):655]." *Psychoneuroendocrinology* 22(3): 189–212.

12. McQueen, J. K., H. Wilson, et al. (1997). "Estradiol-17 beta increases serotonin transporter (SERT) mRNA levels and the density of SERT-binding sites in female rat brain." *Brain Res Mol Brain Res* 45(1): 13–23.

13. Fink, G., B. E. Sumner, et al. (1996). "Estrogen control of central neurotransmission: effect on mood, mental state, and memory." *Cell Mol Neurobiol* 16(3): 325–44.

14. Stahl, S. M. (1998). "Basic psychopharmacology of antidepressants, part 2: Estrogen as an adjunct to antidepressant treatment." *J Clin Psychiatry* 59(Suppl 4): 15–24.

15. Schneider, L. S., G. W. Small, et al. (1997). "Estrogen replacement and response to fluoxetine in a multicenter geriatric depression trial. Fluoxetine Collaborative Study Group." *Am J Geriatr Psychiatry* 5(2): 97–106.

16. Seidman, S. N. and J. G. Rabkin (1998). "Testosterone replacement therapy for hypogonadal men with SSRI- refractory depression." *J Affect Disord* 48(2–3): 157–61.

17. Zweifel and O'Brien. "A meta-analysis. . . ."

18. Haggerty, J. J., Jr. and A. J. Prange, Jr. (1995). "Borderline hypothyroidism and depression." *Annu Rev Med* 46: 37–46.

19. Ibid.

20. Bauer, M. S., P. C. Whybrow, et al. (1990). "Rapid cycling bipolar affective disorder. I. Association with grade I hypothyroidism." *Arch Gen Psychiatry* 47(5): 427–32.

21. Bauer, M., R. Hellweg, et al. (1998). "[High dosage thyroxine treatment in therapy and prevention refractory patients with affective psychoses]." *Nervenarzt* 69(11): 1019–22.

22. Longcope. "Hormone dynamics at the menopause."

23. Gray, A., H. A. Feldman, et al. (1991). "Age, disease, and changing sex hormone levels in middle-aged men: results of the Massachusetts Male Aging Study." *J Clin Endocrinol Metab* 73(5): 1016–25.

24. Barrett-Connor, E., D. G. Von Muhlen, et al. (1999). "Bioavailable testosterone and depressed mood in older men: the Rancho Bernardo Study." *J Clin Endocrinol Metab* 84(2): 573–77.

25. Nishizawa, S., C. Benkelfat, et al. (1997). "Differences between males and females in rates of serotonin synthesis in human brain [see comments]." *Proc Natl Acad Sci U S A* 94(10): 5308–13.

26. Sumner, B. E. and G. Fink (1998). "Testosterone as well as estrogen increases serotonin2A receptor mRNA and binding site densities in the male rat brain." *Brain Res Mol Brain Res* 59(2): 205–14.

27. Vogel, W., E. L. Klaiber, et al. (1985). "A comparison of the antidepressant effects of a synthetic androgen (mesterolone) and amitriptyline in depressed men." *J Clin Psychiatry* 46(1): 6–8.

28. Vogel, W., Klaiber, E. L., Broverman, D. M. (1978). "Roles of the Gonadal Steroid Hormones in Psychiatric Depression in Men and Women." *Progress in Neuro-Psychopharmacology* 2(4): 487–503.

3. Bolstering Brain Power: Memory and Mental Sharpness

1. Caldwell B. M., W. R. I. (1952). "An Evaluation of Psychologic Effects of Sex Hormone Administration in Aged Women: Results of Therapy After Six Months." *J Gerontol* 7: 228–44.

2. Campbell, S. and M. Whitehead (1977). "Oestrogen therapy and the menopausal syndrome." *Clin Obstet Gynaecol* 4(1): 31–47.

3. Sherwin, B. B. (1988). "Estrogen and/or androgen replacement therapy and cognitive functioning in surgically menopausal women." *Psychoneuroendrocrinology* 13(4): 345–57. Phillips, S. M. and B. B. Sherwin (1992). "Effects of estrogen on memory function in surgically menopausal women." *Psychoneuroendrocrinology* 17(5): 485–95. Sherwin, B. B. (1994). "Estrogenic effects on memory in women." *Ann N Y Acad Sci* 743: 213–30; discussion 230–31.

4. Sherwin. "Estrogen and/or androgen replacement therapy . . ."

5. Resnick, S. M., E. J. Metter, et al. (1997). "Estrogen replacement therapy and

longitudinal decline in visual memory. A possible protective effect?" *Neurology* 49(6): 1491–97.

6. Delange, F. (2000). "The role of iodine in brain development." *Proc Nutr Soc* 59(1): 75–79.

7. Monzani, F., P. Del Guerra, et al. (1993). "Subclinical hypothyroidism: neurobehavioral features and beneficial effect of L-thyroxine treatment." *Clin Investig* 71(5): 367–71.

8. Madeira, M. D., N. Sousa, et al. (1992). "Selective vulnerability of the hippocampal pyramidal neurons to hypothyroidism in male and female rats." *J Comp Neurol* 322(4): 501–18.

9. Broverman, D. M., Klaiber, E. L., Vogel W. (2000). "The Effect of Hormone Replacement on a Decision Making Task." Forthcoming.

10. Lawrence, M. (1998). *Menopause and Madness.* Kansas City, Missouri, Andrews McMeel Universal Company.

4. Harnessing the Chemistry of Desire: Sexual Health

1. Rako, S. (1999). *The Hormone of Desire: The Truth about Testosterone, Sexuality, and Menopause.* New York, N.Y., Three Rivers Press.

2. Sarrel, P. M. (1990). "Sexuality and menopause." *Obstet Gynecol* 75(4 Suppl): 26S–30S; discussion 31S–35S.

3. Persky, H., L. Dreisbach, et al. (1982). "The relation of plasma androgen levels to sexual behaviors and attitudes of women." *Psychosom Med* 44(4): 305–19.

4. Panzica, G. C., C. Viglietti-Panzica, et al. (1996). "The sexually dimorphic medial preoptic nucleus of quail: a key brain area mediating steroid action on male sexual behavior." *Front Neuroendocrinol* 17(1): 51–125.

5. Simon, J., E. Klaiber, et al. (1999). "Differential effects of estrogen-androgen and estrogen-only therapy on vasomotor symptoms, gonadotropin secretion, and endogenous androgen bioavailability in postmenopausal women." *Menopause* 6(2): 138–46.

6. Ibid.

7. Alexander, G. M., and B. B. Sherwin (1993). "Sex steroids, sexual behavior, and selection attention for erotic stimuli in women using oral contraceptives." *Psychoneuroendocrinology* 18(2): 91–102.

8. Sherwin, B. B., M. M. Gelfand, et al. (1985). "Androgen enhances sexual motivation in females: a prospective, crossover study of sex steroid administration in the surgical menopause." *Psychosom Med* 47(4): 339–51.

9. Sherwin, B. B. and M. M. Gelfand (1987). "The role of androgen in the maintenance of sexual functioning in oophorectomized women." *Psychosom Med* 49(4): 397–409.

10. Sarrel, P., B. Dobay, et al. (1998). "Estrogen and estrogen-androgen replacement in postmenopausal women dissatisfied with estrogen-only therapy. Sexual behavior and neuroendocrine responses." *J Reprod Med* 43(10): 847–56.

11. Sherwin, B. B. and M. M. Gelfand (1985). "Sex steroids and affect in the surgical menopause: a double-blind, cross-over study." *Psychoneuroendocrinology* 10(3): 325–35.

12. Vogel, W., Klaiber, E. L., Broverman, D. M. (1978). "Roles of the Gonadal Steroid Hormones in Psychiatric Depression in Men and Women." *Progress in Neuro-Psychopharmacology* 2(4): 487–503.

13. Behre, H. M., O. F., Nieschlag, E. (1990). Comparative pharmacokinetics of androgen preparations: application of computer analysis and stimulation. *Testosterone: Action, Deficiency, Substitution.* Berlin, Germany, Springer-Verlag: 115–35.

14. Shifren, J. L., G. D. Braunstein, et al. (2000). "Transdermal Testosterone Treatment in Women with Impaired Sexual Function after Oophorectomy." *N Engl J Med* 343(10): 682–88.

15. Hoyert, D. L., K. K. D., Murphy S. L., (1999). Deaths: final data for 1997. National Statistics Reports, Hyattsville, Md: National Center for Health Statistics. 47: 1–15, 27–37.

16. Hammond, C. (1998). *Therapeutic Options For Menopausal Health—Monograph.* Durham, NC, Duke University Medical Center and MBK Associates, LLC.

17. Honoré, E. K., et al. (1996). "Methyltestosterone does not diminish the beneficial effects of estrogen replacement therapy on coronary artery reactivity in cynomolgus monkeys." *Menopause* 3(1): 20–26.

18. Sherwin, B. B., M. M. Gelfand, et al. (1987). "Postmenopausal estrogen and androgen replacement and lipoprotein lipid concentrations." *Am J Obstet Gynecol* 156(2): 414–19.

19. Raisz, L. G., B. Wiita, et al. (1996). "Comparison of the effects of estrogen alone and estrogen plus androgen on biochemical markers of bone formation and resorption in postmenopausal women." *J Clin Endocrinol Metab* 81(1): 37–43.

20. Phillips, G. B., B. H. Pinkernell, et al. (1997). "Relationship between serum sex hormones and coronary artery disease in postmenopausal women." *Arterioscler Thromb Vasc Biol* 17(4): 695–701.

21. Schiavi, R. C., D. White, et al. (1997). "Effect of testosterone administration on sexual behavior and mood in men with erectile dysfunction." *Arch Sex Behav* 26(3): 231–41.

22. Morales, A., B. Johnston, et al. (1997). "Testosterone supplementation for hypogonadal impotence: assessment of biochemical measures and therapeutic outcomes." *J Urol* 157(3): 849–54.

23. Marin R., E., A., Abreu, P., Mas, M. (1999). "Androgen-dependent nitric oxide release in rat penis correlates with levels of constitute nitric oxide synthase isoenzymes." *Biol Reprod* 61(4): 1012–16.

5. Calming the Storms: The New PMS Prescription

1. Frank, R. T. (1931). "Hormonal Causes of Premenstrual Tension." *Arch Neurol Psychiatry* 26: 1053–57.

2. Singh, B. B., B. M. Berman, et al. (1998). "Incidence of premenstrual syndrome and remedy usage: a national probability sample study." *Altern Ther Health Med* 4(3): 75–79; Campbell, E. M., D. Peterkin, et al. (1997). "Premenstrual symptoms in general practice patients. Prevalence and treatment." *J Reprod Med* 42(10): 637–46.

3. De la Gandara Martin, J. J. and E. de Diego Herrero (1996). "[Premenstrual dysphoric disorder: an epidemiological study]." *Actas Luso Esp Neurol Psiquiatr Cienc Afines* 24(3): 111–17.

4. Dalton, K. (1977). *The Premenstrual Syndrome and Progesterone Therapy.* Chicago, Year Book Medical Publishers Inc.

5. Vliet, E. L. (1995). "Depression, PMS or Perimenopause?: Hidden Hormonal Links." *Screaming to Be Heard : Hormonal Connections Women Suspect . . . and Doctors Ignore.* New York, N.Y., M. Evans and Company, Inc.: 125–64.

6. Brown, C. S., F. W. Ling, et al. (1994). "Efficacy of depot leuprolide in premenstrual syndrome: effect of symptom severity and type in a controlled trial." *Obstet Gynecol* 84(5): 779–86.

7. Klaiber, E. L., D. M. Broverman, et al. (1996). "Individual differences in changes in mood and platelet monoamine oxidase (MAO) activity during hormonal replacement therapy in menopausal women." *Psychoneuroendocrinology* 21(7): 575–92.

8. Vliet, "Depression, PMS or Perimenopause?"

9. Schmidt, P. J., L. K. Nieman, et al. (1998). "Differential behavioral effects of gonadal steroids in women with and in those without premenstrual syndrome [see comments]." *N Engl J Med* 338(4): 209–16.

10. Ashby, C. R., Jr., L. A. Carr, et al. (1988). "Alteration of platelet serotonergic mechanisms and monoamine oxidase activity in premenstrual syndrome." *Biol Psychiatry* 24(2): 225–33; Rapkin, A. J., E. Edelmuth, et al. (1987). "Whole-blood serotonin in premenstrual syndrome." *Obstet Gynecol* 70(4): 533–37.

11. Steiner, M., S. Steinberg, et al. (1995). "Fluoxetine in the treatment of premenstrual dysphoria. Canadian Fluoxetine/Premenstrual Dysphoria Collaborative Study Group [see comments]." *N Engl J Med* 332(23): 1529–34.

12. Watson, N. R., J. W. Studd, et al. (1990). "The long-term effects of estradiol implant therapy for the treatment of premenstrual syndrome." *Gynecol Endocrinol* 4(2): 99–107.

13. Geber, S. and J. P. Caetano (1996). "Doppler colour flow analysis of uterine and ovarian arteries prior to and after surgery for tubal sterilization: a prospective study." *Hum Reprod* 11(6): 1195–98.

14. Brown, Ling, et al. "Efficacy of depot leuprolide. . . ."

15. Goodale, I. L., A. D. Domar, et al. (1990). "Alleviation of premenstrual syndrome symptoms with the relaxation response." *Obstet Gynecol* 75(4): 649–55.

16. Blake, F., P. Salkovskis, et al. (1998). "Cognitive therapy for premenstrual syndrome: a controlled trial." *J Psychosom Res* 45(4): 307–18.

17. Abraham, G. E. and M. M. Lubran (1981). "Serum and red cell magnesium levels in patients with premenstrual tension." *Am J Clin Nutr* 34(11): 2364–66.

18. Abraham, G. E. (1983). "Nutritional factors in the etiology of the premenstrual tension syndromes." *J Reprod Med* 28(7): 446–64.

19. Abraham, G. E. (1986). *Management of the Premenstrual Tension Syndromes: Rationale for a Nutritional Approach.* New Canaan, Connecticut, Keats Publishing, Inc.

20. Bassler, K. H. (1989). "Use and abuse of high dosages of vitamin B6." *Int J Vitam Nutr Res Suppl* 30: 120–26.

21. Maskall, D. D., R. W. Lam, et al. (1997). "Seasonality of symptoms in women with late luteal phase dysphoric disorder." *Am J Psychiatry* 154(10): 1436–41.

22. Parry, B. L., A. M. Mahan, et al. (1993). "Light therapy of late luteal phase dysphoric disorder: an extended study." *Am J Psychiatry* 150(9): 1417–19.

6. The Gift of Consciousness: Estrogen and Alzheimer's Disease

1. Andersen, K., L. J. Launer, et al. (1999). "Gender differences in the incidence of AD and vascular dementia: The EURODEM Studies. EURODEM Incidence Research Group." *Neurology* 53(9): 1992–97.

2. Birge, S. J. and K. F. Mortel (1997). "Estrogen and the treatment of Alzheimer's disease." *Am J Med* 103(3A): 36S–45S; Henderson, V. W., L. Watt, et al. (1996). "Cognitive skills associated with estrogen replacement in women with Alzheimer's disease." *Psychoneuroendocrinology* 21(4): 421–30.

3. Henderson, V. W., A. Paganini-Hill, et al. (2000). "Estrogen for Alzheimer's disease in women: randomized, double-blind, placebo-controlled trial." *Neurology* 54(2): 295–301.

4. Mulnard, R. A., C. W. Cotman, et al. (2000). "Estrogen replacement therapy for treatment of mild to moderate Alzheimer disease: a randomized controlled trial. Alzheimer's Disease Cooperative Study [see comments]." *Jama* 283(8): 1007–15.

5. Wang, P. N., S. Q. Liao, et al. (2000). "Effects of estrogen on cognition, mood, and cerebral blood flow in AD: A controlled study [In Process Citation]." *Neurology* 54(11): 2061–66.

6. Carlson, L. E., B. B. Sherwin, et al. (2000). "Relationships between mood and estradiol (E2) levels in Alzheimer's disease (AD) patients." *J Gerontol B Psychol Sci Soc Sci* 55(1): 47–53.

7. Diaz Brinton, R. and R. S. Yamazaki (1998). "Advances and challenges in the prevention and treatment of Alzheimer's disease." *Pharm Res* 15(3): 386–98.

8. Ibid.

9. Birge, S. J. and K. F. Mortel (1997). "Estrogen and the treatment of Alzheimer's disease." *Am J Med* 103(3a).

10. Prencipe, M., A. R. Casini, et al. (1996). "Prevalence of dementia in an elderly rural population: effects of age, sex, and education." *J Neurol Neurosurg Psychiatry* 60(6): 628–33.

11. Lautenschlager, N., A. Kurz, et al. (1999). "[Inheritable causes and risk factors of Alzheimer's disease]." *Nervenarzt* 70(3): 195–205.

12. Farrer, L. A., L. A. Cupples, et al. (1997). "Effects of age, sex, and ethnicity on the association between apolipoprotein E genotype and Alzheimer disease. A meta-analysis. APOE and Alzheimer Disease Meta Analysis Consortium [see comments]." *Jama* 278(16): 1349–56.

13. Imran, M. B., R. Kawashima, et al. (1999). "Parametric mapping of cerebral blood flow deficits in Alzheimer's disease: a SPECT study using HMPAO and image standardization technique." *J Nucl Med* 40(2): 244–49.

14. Henderson, V. W. (1997). "The epidemiology of estrogen replacement therapy and Alzheimer's disease." *Neurology* 48(5 Suppl 7): S27–35. Ott, A., R. P. Stolk, et al. (1999). "Diabetes mellitus and the risk of dementia: The Rotterdam Study [see comments]." *Neurology* 53(9): 1937–42; Finch, C. E. and D. M. Cohen (1997). "Aging, metabolism, and Alzheimer disease: review and hypotheses." *Exp Neurol* 143(1): 82–102.

15. Henderson, V. W. (1997). "The epidemiology of estrogen replacement therapy and Alzheimer's disease." *Neurology* 48(5 Suppl 7): S27–35.

16. Saxena, P. and J. Shankar (2000). "Contralateral hip fractures—can predisposing factors be determined?" *Injury* 31(6): 421–424; Birge, S. J., N. Morrow-Howell, et al. (1994). "Hip fracture." *Clin Geriatr Med* 10(4): 589–609; Aronson, M. K., W. L. Ooi, et al. (1990). "Women, myocardial infarction, and dementia in the very old [see comments]." *Neurology* 40(7): 1102–6; Berlinger, W. G. and J. F. Potter (1991). "Low Body Mass Index in demented outpatients." *J Am Geriatr Soc* 39(10): 973–78.

17. Stern, Y., B. Gurland, et al. (1994). "Influence of education and occupation on the incidence of Alzheimer's disease [see comments]." *Jama* 271(13): 1004–10; Crawford, J. G. (1998). "Alzheimer's disease risk factors as related to cerebral blood flow: additional evidence." *Med Hypotheses* 50(1): 25–36;

McGeer, P. L., M. Schulzer, et al. (1996). "Arthritis and anti-inflammatory agents as possible protective factors for Alzheimer's disease: a review of 17 epidemiologic studies [see comments]." *Neurology* 47(2): 425–32.

18. Breteler, M. M., J. J. Claus, et al. (1992). "Epidemiology of Alzheimer's disease." *Epidemiol Rev* 14: 59–82. Crawford, J. G. (1996). "Alzheimer's disease risk factors as related to cerebral blood flow." *Med Hypotheses* 46(4): 367–77. Yoshimasu, F., E. Kokmen, et al. (1991). "The association between Alzheimer's disease and thyroid disease in Rochester, Minnesota." *Neurology* 41(11): 1745–47.

19. Mazziotti, M. and D. H. Perlmutter (1998). "Resistance to the apoptotic effect of aggregated amyloid-beta peptide in several different cell types including neuronal- and hepatoma- derived cell lines." *Biochem J* 332(Pt 2): 517–24.

20. Ishiguro, K. (1998). "[Involvement of tau protein kinase in amyloid-beta-induced neurodegeneration]." *Rinsho Byori* 46(10): 1003–7.

21. Henderson, V. W. (1997). "The epidemiology of estrogen replacement therapy and Alzheimer's disease." *Neurology* 48(5 Suppl 7): S27–35.

22. Paganini-Hill, A. and V. W. Henderson (1996). "Estrogen replacement therapy and risk of Alzheimer disease." *Arch Intern Med* 156(19): 2213–17.

23. Ibid.

24. Kawas, C., S. Resnick, et al. (1997). "A prospective study of estrogen replacement therapy and the risk of developing Alzheimer's disease: the Baltimore Longitudinal Study of Aging [published erratum appears in Neurology 1998 Aug;51(2):654]." *Neurology* 48(6): 1517–21.

25. Baldereschi, M., A. Di Carlo, et al. (1998). "Estrogen-replacement therapy and Alzheimer's disease in the Italian Longitudinal Study on Aging." *Neurology* 50(4): 996–1002.

26. Tang, M. X., D. Jacobs, et al. (1996). "Effect of oestrogen during menopause on risk and age at onset of Alzheimer's disease [see comments]." *Lancet* 348(9025): 429–32.

27. Yaffe, K., G. Sawaya, et al. (1998). "Estrogen therapy in postmenopausal women: effects on cognitive function and dementia." *Jama* 279(9): 688–95.

28. Ohkura, T., K. Isse, et al. (1994). "Evaluation of estrogen treatment in female patients with dementia of the Alzheimer type." *Endocr J* 41(4): 361–71.

29. Honjo, H., Y. Ogino, et al. (1989). "In vivo effects by estrone sulfate on the central nervous system-senile dementia (Alzheimer's type)." *J Steroid Biochem* 34(1–6): 521–25.

30. Chang, D., J. Kwan, et al. (1997). "Estrogens influence growth, maturation, and amyloid beta-peptide production in neuroblastoma cells and in a beta-APP transfected kidney 293 cell line." *Adv Exp Med Biol* 429: 261–71.

31. Behl, C., M. Widmann, et al. (1995). "17-beta estradiol protects neurons from oxidative stress-induced cell death in vitro." *Biochem Biophys Res Commun* 216(2): 473–82.

32. Adolfsson, R., C. G. Gottfries, et al. (1980). "Increased activity of brain and platelet monoamine oxidase in dementia of Alzheimer type." *Life Sci* 27(12): 1029–34; Parnetti, L., G. P. Reboldi, et al. (1994). "Platelet MAO-B activity as a marker of behavioural characteristics in dementia disorders." *Aging (Milano)* 6(3): 201–7.

33. Yoshimura, N., Y. Soma, et al. (1993). "[Relations between neuropsychological findings and lateral asymmetries of cerebral blood flow measured by SPECT in dementia of Alzheimer type]." *Rinsho Shinkeigaku* 33(10): 1029–32.

34. Schneider, L. S., M. R. Farlow, et al. (1996). "Effects of estrogen replacement therapy on response to tacrine in patients with Alzheimer's disease." *Neurology* 46(6): 1580–84.

35. Ibid.

36. Small, G. W. (1998). "Treatment of Alzheimer's disease: current approaches and promising developments." *Am J Med* 104(4A): 32S–38S; discussion 39S–42S.

7. The Protected Passage: Hormone Modulation for a Healthy Menopause

1. Simon, J. (1993). Estrogen pharmacokinetics: interindividual variations. *Hormone Replacement Therapy: Standardized or Individually Adapter Doses?* New York, N.Y., The Parthenon Publishing Group.

2. McNagny, S. E., N. K. Wenger, et al. (1997). "Personal use of postmenopausal hormone replacement therapy by women physicians in the United States [see comments]." *Ann Intern Med* 127(12): 1093–96.

3. Brinton, R. D., P. Proffitt, et al. (1997). "Equilin, a principal component of the estrogen replacement therapy Premarin, increases the growth of cortical neurons via an NMDA receptor- dependent mechanism." *Exp Neurol* 147(2): 211–20.

4. Casper, R. F., N. J. MacLusky, et al. (1996). "Rationale for estrogen with interrupted progestin as a new low-dose hormonal replacement therapy." *J Soc Gynecol Investig* 3(5): 225–34; Sulak, P. J., P. Caubel, et al. (1999). "Efficacy and safety of a constant-estrogen, pulsed-progestin regimen in hormone replacement therapy." *Int J Fertil Womens Med* 44(6): 286–96.

5. Klaiber, E. L., D. M. Broverman, et al. (1997). "Relationships of serum estradiol levels, menopausal duration, and mood during hormonal replacement therapy." *Psychoneuroendocrinology* 22(7): 549–58.

6. Utian, W. H. (1989). "Consensus statement on progestin use in postmenopausal women." *Maturitas* 11(3): 175–77.

7. Klaiber, E. L., D. M. Broverman, et al. (1996). "Individual differences in changes in mood and platelet monoamine oxidase (MAO) activity during hormonal replacement therapy in menopausal women." *Psychoneuroendocrinology* 21(7): 575–92.

8. Ibid.

9. Sherwin, B. B. (1991). "The impact of different doses of estrogen and progestin on mood and sexual behavior in postmenopausal women." *J Clin Endocrinol Metab* 72(2): 336–43.

10. Ziel, H. K. and W. D. Finkle (1975). "Increased risk of endometrial carcinoma among users of conjugated estrogens." *N Engl J Med* 293(23): 1167–70.

11. Hulley, S., D. Grady, et al. (1998). "Randomized trial of estrogen plus progestin for secondary prevention of coronary heart disease in postmenopausal women. Heart and Estrogen/progestin Replacement Study (HERS) Research Group [see comments]." *Jama* 280(7): 605–13.

12. Adams, M. R., T. C. Register, et al. (1997). "Medroxyprogesterone acetate antagonizes inhibitory effects of conjugated equine estrogens on coronary artery atherosclerosis." *Arterioscler Thromb Vasc Biol* 17(1): 217–21.

13. Adams, M. R., J. R. Kaplan, et al. (1990). "Inhibition of coronary artery atherosclerosis by 17-beta estradiol in ovariectomized monkeys. Lack of an effect of added progesterone." *Arteriosclerosis* 10(6): 1051–57.

14. Herrington, D. M., D. M. Reboussin, et al. (2000). "Effects of estrogen replacement on the progression of coronary-artery atherosclerosis." *N Engl J Med* 343(8): 522–29.

15. (1995). "Effects of estrogen or estrogen/progestin regimens on heart disease risk factors in postmenopausal women. The Postmenopausal Estrogen/Progestin Interventions (PEPI) Trial. The Writing Group for the PEPI Trial [see comments] [published erratum appears in JAMA 1995 Dec 6;274(21):1676]." *Jama* 273(3): 199–208.

16. Hu, F. B., M. J. Stampfer, et al. (2000). "Trends in the incidence of coronary heart disease and changes in diet and lifestyle in women." *N Engl J Med* 343(8): 530–37.

17. Chaouloff, F. (1989). "Physical exercise and brain monoamines: a review." *Acta Physiol Scand* 137(1): 1–13.

18. McTiernan, A., J. L. Stanford, et al. (1998). "Prevalence and correlates of recreational physical activity in women aged 50–64 years." *Menopause* 5(2): 95–101.

19. Stefanick, M. L., S. Mackey, et al. (1998). "Effects of diet and exercise in men and postmenopausal women with low levels of HDL cholesterol and high levels of LDL cholesterol [see comments]." *N Engl J Med* 339(1): 12–20.

20. Murkies, A. L., C. Lombard, et al. (1995). "Dietary flour supplementation decreases post-menopausal hot flushes: effect of soy and wheat." *Maturitas* 21(3): 189–95.

21. Baird, D. D., D. M. Umbach, et al. (1995). "Dietary intervention study to assess estrogenicity of dietary soy among postmenopausal women." *J Clin Endocrinol Metab* 80(5): 1685–90.

22. Wheatley, D. (1997). "LI 160, an extract of St. John's wort, versus amitriptyline in mildly to moderately depressed outpatients—a controlled 6-week clinical trial." *Pharmacopsychiatry* 30 Suppl 2: 77–80.

23. Rako, S. (1996). *The Hormone of Desire: The Truth about Testosterone, Sexuality, and Menopause.* New York, N.Y., Crown Publishing Group.

9. The Safety Equation: Weighing Risks and Benefits

1. Gapstur, S. M., M. Morrow, et al. (1999). "Hormone replacement therapy and risk of breast cancer with a favorable histology: results of the Iowa Women's Health Study [see comments]." *Jama* 281(22): 2091–97.

2. Schairer, C., J. Lubin, et al. (2000). "Menopausal estrogen and estrogen-progestin replacement therapy and breast cancer risk [see comments]." *Jama* 283(4): 485–91; Helzlsouer, K. J. and R. Couzi (1995). "Hormones and breast cancer." *Cancer* 76(10 Suppl): 2059–63.

3. Hammond, C. (1998). *Therapeutic Options For Menopausal Health—Monograph.* Durham, N.C., Duke University Medical Center and MBK Associates, LLC.

4. Ibid.

5. Grodstein, F., M. J. Stampfer, et al. (1997). "Postmenopausal hormone therapy and mortality [see comments]." *N Engl J Med* 336(25): 1769–75.

6. Coelho, R., C. Silva, et al. (1999). "Bone mineral density and depression: a community study in women." *J Psychosom Res* 46(1): 29–35.

7. Yaffe, K., W. Browner, et al. (1999). "Association between bone mineral density and cognitive decline in older women." *J Am Geriatr Soc* 47(10): 1176–82.

8. Devogelaer, J. P., H. Broll, et al. (1996). "Oral alendronate induces progressive increases in bone mass of the spine, hip, and total body over 3 years in postmenopausal women with osteoporosis [published erratum appears in *Bone* 1996 Jul;19(1):78]." *Bone* 18(2): 141–50.

9. Ravn, P., M. Bidstrup, et al. (1999). "Alendronate and estrogen-progestin in the long-term prevention of bone loss: four-year results from the early postmenopausal intervention cohort study. A randomized, controlled trial." *Ann Intern Med* 131(12): 935–42. Bone, H. G., S. L. Greenspan, et al. (2000). "Alendronate and estrogen effects in postmenopausal women with low bone mineral density. Alendronate/Estrogen Study Group." *J Clin Endocrinol Metab* 85(2): 720–26. Lindsay, R., F. Cosman, et al. (1999). "Addition of alendronate to ongoing hormone replacement therapy in the treatment of osteoporosis: a randomized, controlled clinical trial." *J Clin Endocrinol Metab* 84(9): 3076–81.

10. Taking Charge of Your Hormones: The Decision Is Yours

1. Marvel, M. K., R. M. Epstein, et al. (1999). "Soliciting the patient's agenda: have we improved? [see comments]." *Jama* 281(3): 283–87.

11. New Medicine for the Mind: The Future of Hormone Modulation

1. Klaiber, E. L., D. M. Broverman, et al. (1979). "Estrogen therapy for severe persistent depressions in women." *Arch Gen Psychiatry* 36(5): 550–54.

2. Stahl, S. M. (1998). "Basic psychopharmacology of antidepressants, part 2: Estrogen as an adjunct to antidepressant treatment." *J Clin Psychiatry* 59(Suppl 4): 15–24.

3. Hammond, C. (1998). *Therapeutic Options For Menopausal Health—Monograph.* Durham, N.C., Duke University Medical Center and MBK Associates, LLC; Paganini-Hill, A. and V. W. Henderson (1996). "Estrogen replacement therapy and risk of Alzheimer disease." *Arch Intern Med* 156(19): 2213–17.

4. Schneider, L. S., M. R. Farlow, et al. (1996). "Effects of estrogen replacement therapy on response to tacrine in patients with Alzheimer's disease." *Neurology* 46(6): 1580–84.

5. Saunders-Pullman, R., J. Gordon-Elliott, et al. (1999). "The effect of estrogen replacement on early Parkinson's disease." *Neurology* 52(7): 1417–21; Marder, K., M. X. Tang, et al. (1998). "Postmenopausal estrogen use and Parkinson's disease with and without dementia." *Neurology* 50(4): 1141–43.

Glossary

Acetylcholine: A neurotransmitter important to normal memory function, which is deficient in Alzheimer's disease.

Adrenal glands: Two small glands sitting atop each kidney, which secrete cortisol, DHEA, and androstenedione, which are not strong androgens themselves but can be converted to the potent androgen testosterone. The adrenals also secrete estradiol and estrone in small amounts, which are minimal compared to the amounts secreted by the ovary. Larger amounts of estrogens are derived from the adrenals indirectly by conversion of androstenedione to estradiol in peripheral tissues such as fat and muscle.

Adrenergic: Refers to adrenergic nerve fibers that secrete the neurotransmitter norepinephrine.

Alendronate (Fosamax): A drug used to stimulate bone formation, useful in patients with osteoporosis.

Alzheimer's disease: A condition occurring late in life and worsening with time in which brain cells die; it is accompanied by memory loss, confusion, and physical decline. There are indications that an estrogen deficiency may predispose to Alzheimer's disease.

Androgens: Sex hormones, such as testosterone, secreted by the testicle, ovary, and adrenal gland which influences sex drive, muscle development, energy and mood. High levels in women may result in excessive body and facial hair, along with possible balding and acne.

Andropause: Andropause, a syndrome in older men, includes physical, sexual, and psychologic symptoms that consist of weakness, fatigue, reduced muscle and bone mass, sexual dysfunction, depression, anxiety, irritability, insomnia, memory impairment, and reduced cognitive function.

Androstenedione: An androgenic hormone produced by the testicle, adrenal and ovary, which can be converted in peripheral tissues to testosterone and estrogen.

Atherosclerosis: A narrowing of the lining of the arteries due to the accumulation of fat and other materials, which may produce constriction of the coronary arteries, resulting in heart disease. Narrowing of arteries in the brain may lead to stroke and other disorders.

Autoimmune thyroiditis: A slowly developing chronic inflammation of the thyroid gland that frequently results in hypothyroidism, a condition of decreased function of the thyroid gland. It is caused by a reaction of the immune system producing thyroid antibodies, which attack the thyroid gland, reducing its function.

Axon: A prolonged tubular projection of a nerve with branches at its end, which transmits signals to other nerve cells. See figure 1.

Bipolar disorder: A mood disorder in which there are mood swings from a feeling of excessive well-being to depression.

Biopsy: The surgical removal and microscopic examination of such tissues as endometrium or breast, to establish a specific diagnosis such as cancer.

Bone density: Bone density is the amount of bone tissue in a certain volume of bone. It can be measured using special X-ray techniques. A low bone density is characteristic of osteoporosis.

Calcitonin: A hormone from the parathyroid glands, which inhibits the breakdown of bone. It can be given as a medication (Miacalcin) to rebuild bone in patients with osteoporosis.

Cholinergic: A type of nerve cell that secretes acetylcholine, which is important in memory and mood states.

Cognition: The operation of the mind that includes all aspects of perceiving, thinking, and remembering.

Cortisol: A hormone produced by the adrenal gland important in the metabolism of sugars, proteins, fats, and salt. It also has can be used as a medication to reduce inflammation.

Dendrite: A treelike projection of a nerve cell that receives messages from other neurons.

DHEA (dehydroepiandrosterone): An adrenal hormone classified as an androgen, which can be converted to estradiol and testosterone.

DHEAS (dehydroepiandrosterone sulfate): It is a sulfur containing salt of DHEA, which can be converted to estradiol and testosterone.

Dopamine: A neurotransmitter found in the brain, important in mood states and schizophrenia.

Double-blind study: A kind of clinical study in which neither the participants nor the person administering treatment know which treatment any particular subject is receiving.

Down's syndrome: A fairly common type of mental retardation in which there are a characteristic face and multiple malformations. Down's syndrome is secondary to a chromosomal abnormality.

Endocrinology: The study of the diseases and medical conditions of hormones produced by specific endocrine glands that secrete their products into the blood stream.

Endometrium: The inner layer of the womb or uterus. This tissue is shed monthly in response to the hormonal changes of the menstrual period.

ERT (Estrogen Replacement Therapy): Estrogen replacement therapy is used to treat menopause. It reduces or stops the short-term changes of menopause such as hot flashes, disturbed sleep, and vaginal dryness.

Estradiol: One of the three major estrogens, which include estrone and estriol. Estradiol is by far the most biologically active of the three.

Estrogen: A generic term for the three major estrogens, estradiol, estriol, and estrone. Estrogens are hormones produced by the ovaries and adrenals responsible for female secondary sex characteristics such as breasts.

Estrogen receptor: A protein on or near the surface of estrogen's sensitive cells to which the hormone must attach in order to express itself.

Estrone: An estrogenic hormone secreted by the ovary and adrenal which is less potent then estradiol but more so than estriol. It can be converted to estradiol in the body.

Fibromyalgia: Also known as fibrositis, fibromyalgia chronically causes pain, stiffness, and tenderness of muscles, tendons, and joints without detectable inflammation.

Follicle Stimulating Hormone (FSH): A hormone secreted by the pituitary gland, which stimulates the ovarian follicles to mature and secrete estrogen. It becomes elevated after menopause. In the male it stimulates sperm production.

Follicular phase: That phase of the menstrual cycle, which precedes ovulation and is associated with the development of the estrogen secreting ovarian follicle which contains the egg.

Free testosterone: The unbound physiological active form of the hormone; the vast majority of testosterone is bound to a blood protein and is inactive.

Estrogen increases the protein binding of testosterone and lowers blood free testosterone levels.

Free thyroxine (free T4): The unbound form of the thyroid hormone thyroxine (T4). The great majority of thyroxine is bound to a blood protein and is physiologically inactive. Estrogen increases the protein binding of thyroxine and lowers blood free thyroxine levels.

GABA (gamma-amino butyric acid): A neurotransmitter that has some anti-anxiety action. Its levels are stimulated by progesterone.

Graves' disease: A common cause of hyperthyroidism (too much thyroid hormone), most often found in women. Graves' disease is due to a generalized (diffuse) overactivity (toxic) of the whole enlarged thyroid gland (goiter).

Hormone: A chemical substance produced in the body that controls and regulates the activity of certain cells and organs. Hormones are necessary for every function of daily living, including the processes of digestion, metabolism, growth, reproduction, and mood control.

Hot flash: A sudden wave of intense body heat caused by rushes of hormonal changes and triggered by estrogen deficiency.

HRT: (Hormone Replacement Therapy): The combination therapy of estrogen plus progestin, used to treat menopause. The progestin can be given either daily with the estrogen or in a cyclic fashion for 10 to 14 days each month.

Hyperplasia: A condition in which there is an increase in the number of normal cells in a tissue or organ, which may occur in the endometrium sometimes associated with estrogen replacement therapy. If this condition progresses it can eventually lead to uterine cancer.

Hyperthyroidism: Excess of thyroid hormone resulting from an overactive thyroid gland (or taking too much thyroid hormone).

Hypothalamus: The area of the brain that controls body temperature, hunger, and thirst. It also activates the pituitary to produce its trophic hormones.

Hypomania: An abnormality of mood resembling mania (unrestrained elevated mood) but of lesser intensity.

Hypothyroidism: Deficiency of thyroid hormone, which is made by the thyroid gland, which is located in the front of the neck.

Hysterectomy: An operation to remove the uterus and sometimes the cervix.

Intramuscular: An intramuscular (IM) medication is given by needle into the muscle.

Luteal phase: The postovulatory phase of a woman's cycle.

Lutenizing Hormone (LH): A hormone released by the pituitary gland. It controls the length and sequence of the female menstrual cycle.

MAO (monoamine oxidase) inhibitors: A class of antidepressant medications which inhibit the levels of monoamine oxidase in the brain.

Mania: An abnormally elevated mood condition characterized by such symptoms as excessive elation, increased irritability, insomnia, grandiose notions, and increased speed and/or volume of speech.

Mastalgia: Pain in the breast or mammary gland.

Medroxyprogesterone: A synthetic progesterone used in hormone replacement therapy, to treat menopause.

Menopause: The time in a woman's life when menstrual periods permanently stop, characterized by symptoms of estrogen and testosterone deficiency, including hot flashes, night sweats, vaginal dryness, decreased sex drive, and mood disorders.

Methyltestosterone: A synthetic androgen having actions similar to those of testosterone. It is used as a replacement therapy for testosterone deficiency in both men and women, and usually administered orally.

Micronized progesterone: Micronization of progesterone is a process by which the synthesized progesterone is ground to extremely fine particles (micronized) that can be more easily absorbed through the gastrointestinal tract.

Monoamine oxidase (MAO): An enzyme found in nerve cells which inactivates neurotransmitters which have a single or mono-amine group in their chemical structure. High or low levels are observed in mood disorders.

Neurology: A branch of medical science that deals with the nervous system, both normal and in disease.

Neurotransmitter: A chemical that is secreted from a nerve cell which transmits an impulse from one nerve cell to another, muscle, organ, or other tissue. A neurotransmitter is a messenger of neurologic information from one cell to another, which may have many important effects on mood.

Neuron: A nerve cell. Neurons send and receive electrical and chemical signals over long distances in the body. See Figure 1.

Norepinephrine: A chemical neurotransmitter released by adrenergic nerve fibers in the brain and elsewhere in the nervous system. A deficiency of norepinephrine may result in depression and excess may cause anxiety.

Osteoporosis: Thinning of the bones with reduction in bone density due to depletion of calcium and bone protein. Osteoporosis predisposes a person to fractures after minimal trauma. It is more common in older adults, particularly postmenopausal women and is secondary to estrogen deficiency. Osteoporosis can be prevented by estrogen replacement therapy.

Ovary: The ovary is one of a pair of reproductive glands in women. They are located in the pelvis, one on each side of the uterus. Each ovary is approximately the shape and size of an almond. The ovaries produce eggs (ova) and the hormones estrogen, testosterone, and progesterone.

Parkinson's disease: An abnormal condition of the nervous system caused by deterioration of certain areas of the brain, and by low production of the neurotransmitter dopamine. The disease results in rigidity of the muscles, slow body movement, and tremors. New studies are indicating that an estrogen deficiency may predispose to Parkinson's disease.

Perimenopause: A period of time five to six years before menopause when women's menstrual periods frequently begin to change, becoming less regular. This is a time when the levels of hormones produced by the aging ovaries fluctuate widely. Other changes associated with the perimenopause include night sweats, mood swings, vaginal dryness, fluctuations in sexual desire (libido), forgetfulness, trouble sleeping and fatigue, probably from loss of sleep and mood changes. Eventually levels of estrogen decline and levels of FSH begin to increase.

Phytoestrogen: Estrogenlike substances derived from plants such as soy. These have become popular among women who want substitutes for the natural form of estrogen. These substances have only weak estrogenic action and have not been proven to have the therapeutic effects attributed to estrogen itself.

Pituitary gland: The main endocrine gland. It is a small structure at the base of the brain below the hypothalamus. It is called the master gland because it produces hormones that control other endocrine glands such as the ovary, testicle, thyroid, and adrenal. It also influences many body functions including growth.

Pituitary gonadotrophin: The pituitary gonadotrophic hormones which stimulate the ovary and the testicle. The gonadotrophins are follicle stimulating hormone and lutenizing hormone. They are important in controlling ovulation and the production of estrogen, testosterone and progesterone during the menstrual cycle.

Postmenopause: Postmenopause is defined as the time after which a woman has experienced twelve consecutive months without menstruation, characterized by an increase in blood levels of FSH. The average age for the onset of menopause is fifty-one years. In earlier times, life expectancy did not exceed fifty years, and menopause was not considered to be a major problem, but the average length of the postmenopause has been increasing. With greater longevity, a woman will soon be postmenopausal on average a third of her life.

Postpartum: A period of time following pregnancy and delivery, which may be associated within a small percentage of women with a depressive episode (postpartum depression). Such a depression may be in part triggered by a sudden dramatic drop in estrogen levels which are extremely high during pregnancy and then return to normal non-pregnancy levels rapidly after delivery.

Premarin: The most widely prescribed estrogen for the treatment of menopausal symptoms contains a number of estrogenic substances, including estrone sulfate and equiline, which are estrogens derived from the urine of pregnant mares.

Premenstrual syndrome (PMS): A combination of physical and mood disturbances that occur after ovulation and normally end with the onset of the menstrual flow. Premenstrual syndrome is believed to be a disorder of the brain neurotransmitters. In its severest form, it can be truly disabling for part of the month. Treatment of mild PMS may not be necessary, but treatment of moderate to severe PMS may involve exercise, antidepressants and/or mood stabilizers.

Progesterone: An ovarian hormone secreted in large amounts following ovulation. It is administered as a part of HRT and is necessary to prevent the overdevelopment of the endometrium, which may lead to cancer.

Progestin: The name originally given to the crude hormone from the *corpora lutea* of the ovaries. The term is now used for synthetic and naturally occurring progestational agents (substances that have significant activity similar to the hormone progesterone).

Prolactin: A pituitary hormone that stimulates and sustains lactation (milk production from the breasts) following childbirth. A prolactinoma is a pituitary tumor that secretes excessive amounts of prolactin that may interfere with the normal menstrual cycle and decrease sex drive or impair fertility in both sexes.

Proliferative phase: The phase of the menstrual cycle prior to ovulation during which estrogen levels increase and stimulate the proliferation or growth of the endometrium or lining of the uterus. See figure 4.

Psychology: The branch of science that deals with the mind and mental processes, especially in relation to human behavior and interactions.

Psychoneuroendocrinology: A field of medicine dedicated to the understanding of the mind-brain-hormone interconnection.

Psychopharmacologist: A psychiatrist who studies the effects of drugs and medicines on psychological processes and psychiatric diseases, and uses this information to treat patients.

Raloxifene: A "selective estrogen receptor modulator" (more commonly, a SERM), which does not have an estrogen-stimulating effect on the breast and endometrium (uterine lining). It is being prescribed to treat women concerned about the breast and uterine cancer risks of HRT. It does not appear that drugs like raloxifene have the same protective effect that estrogen does against heart disease and neurodegenerations such as Alzheimer's disease. These SERMs do have some protective effects against osteoporosis.

Receptor: In cell biology, a receptor is a structure on the surface of a cell (or inside a cell—see figure 2) that selectively receives and binds a specific substance such as estrogen or serotonin. The receptor is required in order for the hormone or neurotransmitter to carry out its function. Estrogen deficiency results in a decrease in the number of receptors for estradiol and serotonin.

Secretory phase: The phase of the menstrual cycle following ovulation during which estrogen and progesterone levels rise and stimulate the secretion of glands in the endometrium (uterine lining) in preparation for the possible implantation of the fertilized egg, leading to a pregnancy. See figure 4.

Serotonergic: Refers to serotonergic nerve fibers that secrete the neurotransmitter serotonin.

SERMs: A "selective estrogen receptor modulator" is a substance with selective estrogenic activity, which does not stimulate breast tissue or the endometrium but does have some of the same effects of estrogen on other tissues.

Serotonin: A neurotransmitter in the brain involved in the transmission of nerve impulses. A deficiency of serotonin has been implicated in depression.

Sildenafil citrate (Viagra): A drug prescribed for males with erectile dysfunction that helps them to achieve a normal erection.

SSRI: An abbreviation for selective serotonin reuptake inhibitors, commonly prescribed drugs to treat depression. SSRIs affect a neurotransmitter, serotonin, in the brain. Serotonin is released by one nerve and taken up by other nerves. Serotonin that is not taken up by other nerves is taken up by the same nerves that released it in a process termed "reuptake." SSRIs work by inhibiting the reuptake of serotonin, an action which allows more serotonin to be available to be taken up by other nerves. Since serotonin deficiency can cause depression this makes SSRIs useful in the treatment of depression. Prozac is one SSRI.

Tamoxifen: An antiestrogen (a drug that blocks the effects of estrogen), which competes with estrogen for binding sites in target tissues such as breast. It has been approved by the U.S. Food and Drug Administration (FDA) to treat breast cancer and help prevent it in women at high risk for the disease.

Testosterone: A sex hormone produced by the testes, ovaries, and adrenals that encourages the development of male sexual characteristics. However, women also produce testosterone, which is important for normal body hair growth, muscle development, sexual drive and function, and a normal mood. Testosterone is the most potent of the naturally occurring androgens.

Thromboembolic events: A thromboembolism involves clot formation in a vein of the heart. The clot may break off and travel through the circulation to the lungs, or to the brain, causing a stroke. In certain conditions estrogen may predispose to thromboembolic events.

Thrombophlebitis: Inflammation of a vein associated with a clot formation. In certain conditions estrogen may predispose to thrombophlebitis.

Thyroid antibodies: Thyroid peroxidase and thyroglobulin antibodies are produced when thyroid tissue leaks into the blood stream. These antibodies can attack the thyroid gland and impair its function producing hypothyroidism. This condition is referred to as autoimmune thyroiditis and is more commonly found in women immediately postpartum or at the time of menopause.

Thyroxine (T4): A hormone made by the thyroid gland. Thyroid hormones are essential for the function of every cell in the body. They help regulate growth and the rate of chemical reactions (metabolism) in the body and are essential for normal central nervous system function and growth.

Transdermal: A method of administering a hormone utilizing a patch containing the medication applied to the skin. The hormone passes through the skin into the blood stream. Estrogen and testosterone can be administered transdermally.

Tricyclic antidepressants: A type of medication used to treat depression.

The tricyclic antidepressants are also used for some forms of anxiety, fibromyalgia, and to control chronic pain. This type of antidepressant is used less widely since SSRIs have become available. The SSRIs have considerably fewer side effects than the tricyclic antidepressants.

Triglycerides: Triglycerides are fatty acids which circulate in the blood stream. They are synthesized by the liver and this synthesis is elevated in people with high carbohydrate diets. Elevated levels of triglycerides have been implicated in the development of atherosclerosis but their exact relationship to coronary heart disease is undecided. Estrogen administration can elevate triglycerides moderately.

Triiodothyronine (T3): Produced in small quantities by the thyroid gland but most is produced by conversion from thyroxine in the peripheral tissues, chiefly the liver. T3 is more biologically active than thyroxine and is the form of thyroid hormone that is most active in the tissues.

Tubal ligation: A surgical procedure that involves closure of the fallopian tubes to prevent a fertilized egg from reaching the uterus, which is used as a birth control method.

Uterus: The uterus (womb) is a hollow, pear-shaped organ located in a woman's lower abdomen between the bladder and the rectum. The lower portion of the uterus is the cervix. The uterine lining or endometrium is stimulated by estrogen. As a result of falling estrogen levels the lining sheds each month during the menstrual flow.

Virilization: The development in the female of male characteristics such as a deepened voice, an increase in body and facial hair and acne, a decrease in breast size, an enlargement of the clitoris, and "male-pattern" baldness. Virilization can occur secondary to the administration of excessive amounts of testosterone to women.

Index

acetylcholine, 16
  AD and, 75, 154, 158–59, 166–67
  estrogen and, 19, 75–77, 158, 167
acetylcholinesterase (AchE) inhibitors, 159, 162–63, 166
acne, 250
AD, *see* Alzheimer's disease
adrenal glands, 1, 81, 119
  testosterone produced by, 100, 104, 112
aging, *see* elderly
agoraphobia, 139
Alendronate (Fosamax), 244, 246, 253
Alexander, G. M., 108–9
alpha tocopheral (vitamin E), 157, 166–67
alternative medicine, 201–4, 206
Alzheimer, Alois, 169
Alzheimer's disease (AD), 3, 5, 7, 147–71, 201
  blocking development of, 154–60, 162–64, 166–68
  brain and, 10, 13, 19, 21, 149, 152–54, 157–60, 166, 169–70
  delaying onset of, 168
  devastation of, 147–48, 165–66
  diagnosis of, 162, 169, 213–14
  estrogen and, 147, 149–69, 171, 198, 231, 239, 249, 254, 271
  ginkgo biloba and, 204
  hormone modulation and, 169, 202, 208, 213–14, 271, 274
  HRT and, 175, 194
  incidence of, 150–51, 154–55, 163, 167
  memory lapses and, 75, 77, 84–86, 147–48, 151, 154, 158–59, 161–62, 164–67, 169–70
  prevention of, 148–50, 152, 154, 157–58, 166–67, 168, 171, 271
  raloxifene and, 253
  risk factors for, 151–53, 155–56, 159, 164, 166, 168, 171
  symptoms of, 149–50, 156–57, 161, 163, 166, 169–71
  treatments for, 158–63, 166–67
amitriptyline, 55, 66, 112, 203
amygdala, 75–76
andropause, menopause vs., 123
androstenedione, 119–20
anti-inflammatory drugs, 152–53
anxiety, 46–50, 247
  AD and, 161
  estrogen and, 28, 46–47, 50
  hormone modulation and, 181, 185, 188–90, 210–11, 217–18, 221, 223–26, 249
  PMS and, 59, 129, 131–34, 136–39, 141

anxiety (*cont.*):
symptoms of, 46–47
thyroid hormone and, 52–56
ApoE gene, 151, 153
Archives of General Psychiatry, 38–39
Aricept (donezepil), 159–63, 166, 271
autoimmune thyroiditis, 26–27, 53, 86–87, 140, 212, 214, 219
automatization, 72–73, 89–91
axons, 15–16, 76

Barbach, Lonnie, 36–37
beta amyloid, 152–54, 157–58, 167, 169
bipolar disorder, *see* manic-depression
blocking ovulation, 143–44
blood flow:
AD and, 149, 152–53, 157, 159
estrogen and, 20–21, 240
hormone modulation and, 196, 270–71
mood problems and, 45
blood tests, 59, 210–11, 213–15
bone mineral density (BMD), 214, 242–43
brain, 1–2, 5, 111, 121, 259
AD and, 10, 13, 91, 149, 152–54, 157–60, 166, 169–70
blood flow to, *see* blood flow
cognition and, 3–4, 11–22, 25–29, 70, 72–73
exercise and, 200
health of, 274–75
hormone modulation and, 98, 175, 177–79, 183, 185, 194, 197, 204, 206, 208, 213, 220, 261, 265, 269–72, 274–75
and HRT risks and benefits, 230, 234, 243, 245–46, 250
memory and, 9–14, 18, 20–22, 24–29, 75–78, 85, 87
mental sharpness and, 91
mood problems and, 9–29, 41–42, 45, 48, 68, 113
new treatments for, 27–28
PMS and, 130, 132, 134, 140–45
relationship between hormones and, 9–29, 48
sexual health and, 10–14, 22–29, 104–5
testosterone and, 11–12, 15, 18, 21–28, 116, 126
breast cancer:
age and, 237–40
estrogen and, 6–7, 88–89, 232–37, 249–55
hormone modulation and, 176, 181–82, 187, 189, 193, 201–2, 208, 212, 272
and HRT risks and benefits, 229–40, 249–57, 260, 265, 272
mortalities from, 232–33, 239–41, 257
progesterone and, 236–37
Brigham and Women's study, 232–34
Brinton, Louise, 234
British Medical Journal, 203
Brody, Jane, 101–2, 234
Broverman, Donald, 3–4, 15, 18, 37–38, 72, 91–92, 266
Bush, Trudy, 235

CAD (coronary artery disease), 196–201
calcitonin (Miacalcin), 244
cancer, *see specific cancers*
cholesterol:
exercise and, 200
hormone modulation and, 178, 193, 195–97, 202, 214, 223
and HRT risks and benefits, 240–42, 253
testosterone and, 117–18
clitoris, 106–7
Cognex (tacrine), 13, 159–60, 166, 271
cognition, cognitive health, 2–8, 69–78, 143, 201
AD and, 149–50, 154, 157–60, 162, 167

brain and, 3–4, 11–22, 25–29, 70, 72–73
estrogen and, 69–78, 81–82, 89–91, 94–98, 217
ginkgo biloba for, 84, 204
hormone modulation and, 70–71, 73, 94–96, 175, 183, 187–88, 202, 204, 209–10, 213–14, 216–17, 223–26, 228, 259, 261, 264, 267, 270–71, 274–75
HRT and, 71–75, 229–31, 243, 249–50, 252–54
mechanisms of, 15–22, 75–78
in men, 65, 72
progesterone and, 72–73, 94–98, 119
psychological symptoms and, 93–98
testosterone and, 72–73, 95–96, 126
see also concentration; decision-making; memory, memory loss; mental sharpness
Cohen, Jay S., 174–75
Collins, Mary, 23, 266
colon cancer, 7, 230, 237–39
communication:
doctor-doctor, 265–67, 272–73
doctor-patient, 262, 264–65, 272–74
complementary medicine, 204–5
concentration, 1, 69, 71, 79–80, 88–91
estrogen and, 23–24, 27, 80, 89–91
hormone modulation and, 23–24, 95–98, 216, 221–22, 224–25, 228
and HRT risks and benefits, 231, 256
PMS and, 130
coronary artery disease (CAD), 196–201
Cutler, Scott, 139

Dalton, Katharina, 132
decision-making, 1, 3, 5, 69–70, 89–93
AD and, 170
estrogen and, 11, 22, 90–93
hormone modulation and, 11, 22, 73, 95, 98, 212, 216
and HRT risks and benefits, 231
mood problems and, 93
dehydroepiandrosterone (DHEA), 119
dendrites, 15–16, 19, 76
Depakote, 56, 266
Depo-Testosterone (testosterone cypionate), 114, 126
depression, 1–5, 31–47, 74, 110–16, 120–21, 272–73
AD and, 149–50, 152–53, 161
biochemical basis of, 41–44
blood and, 59, 210–11, 213–15
brain and, 11–13, 18–19, 22–24, 26–29, 113
cognition and, 70
definition of, 57
diet and exercise for, 199–201
estrogen and, 4, 32–45, 50, 52, 94–96, 98, 111–13, 216–20, 227–28, 243, 245, 249, 264, 266, 269
factors in development of, 58
hormone modulation and, 4, 28, 57–64, 94–98, 191–93, 210–14, 221–25, 226–28, 261, 269–70
HRT and, 40–41, 49, 175, 177, 181, 185–86, 188, 208, 231, 247, 249, 256
in men, 34, 44, 65–68, 112, 270
PMS and, 34, 58–61, 131–38, 140–41, 145
postpartum, 28, 34–35, 43–45, 53, 56, 58, 61–63, 106, 110–11
progesterone and, 35, 40–41, 47, 49, 95–96, 111, 119, 121
St. John's wort for, 203
sexual health and, 111–12
symptoms of, 57
testosterone and, 41, 52, 111–16, 126, 224–25, 270

depression (*cont.*):
thyroid hormone and, 52–57
treatment-resistant, 22–24, 33, 39, 42–46, 54, 63, 231
diabetes, 152–53, 212
diet, *see* nutrition
dihydrotestosterone (DHT), 119
diuretics, 191
donezepil (Aricept), 159–63, 166, 271
dopamine, 16, 41, 77, 144, 158, 203
Down's syndrome, 151–53
Dr. Susan Love's Breast Book (Love), 257

ED (erectile dysfunction), 123, 125, 127
EEGs (electroencephalograms), 72, 157
elderly:
AD and, 151, 153, 155–56, 170–71
estrogen deficiency in, 237–39, 243–44
memory in, 77–78, 84, 86
mood problems and, 44, 65–66
sexual health and, 127
thyroid hormone in, 52–53
electroencephalograms (EEGs), 72, 157
embolisms, 195, 248
emotional health, 2–4, 6–7, 31–68, 74, 260
AD and, 158–59, 167
diet and exercise for, 199, 201
hormone modulation and, 183, 190–91, 202, 207, 219–21, 224
and HRT risks and benefits, 252, 255
neurotransmitters and, 18
PMS and, 130
and relationship between hormones and brain, 13, 18–21, 28–29
sexual health and, 106, 122–23
testosterone and, 115–16, 120
energy, 28
AD and, 162
diet and exercise for, 199
estrogen and, 122, 189, 191–92, 216–17
hormone modulation and, 121, 191–92, 220–23
HRT and, 177, 245–46, 256
medications and, 212
mood problems and, 57, 62
PMS and, 58–61, 130, 137–38, 141, 145
testosterone and, 120–22, 189, 191–92, 206
thyroid and, 212
equilin, 178–79
erectile dysfunction (ED), 123, 125, 127
ERT, *see* estrogen replacement therapy
Estrace, 64, 217–18
hormone modulation and, 179–80, 224, 227
estradiol, 121, 124
AD and, 157–58
hormone modulation and, 179–81, 184, 211, 213, 215–16, 219, 223–24
mood problems and, 59–61, 64, 66
PMS and, 138
Estratest, 113–14
Estratest H.S., 113–14
estrogen, 1–7, 80–82, 86, 116, 200
AD and, 147, 149–69, 171, 198, 231, 239, 249, 254, 271
administration of, 138–39, 176, 178–81, 217–18
breast cancer and, 6–7, 88–89, 229, 232–37, 249–55
CAD prevented by, 196–99
cognition and, 69–78, 81–82, 89–91, 94–98, 217, 249–50
colon cancer and, 237–39
cyclical changes in, 60
energy and, 121

heart disease and, 5, 7, 116–17, 144, 164, 239–42, 253, 265
hormone modulation and, 173–74, 176–82, 184–89, 191–99, 201–2, 205–6, 210–14, 216–22, 224–28, 259, 264, 266, 269–72
and HRT risks and benefits, 229–31, 232–57, 265–66
memory lapses and, 74–87, 90
for men, 124–27
menopausal decline of, 2
mental sharpness and, 88–93
mood problems and, 32–52, 56–61, 63–67, 77, 82, 93, 95–96, 98, 111–13, 216–18, 220, 249–50
for osteoporosis, 144, 164, 171, 242–46
PMS and, 130–46
relationship between brain and, 10–12, 14–15, 18–29, 48
sexual health and, 51–52, 80, 101–12, 118, 122, 124
types of, 178–81
estrogen/progesterone combinations:
hormone modulation and, 180–81, 188–90, 194–98, 222
risks and benefits of, 236
sexual health and, 118
types of, 180–81
"Estrogen Replacement and Atherosclerosis (ERA)," 242
estrogen replacement therapy (ERT), 266
AD and, 152–57, 166
embolisms and, 248
hormone modulation and, 186–87, 193, 195, 212, 216, 225
memory and, 82
mental sharpness and, 88–89
mood problems and, 40–41, 43–44, 61
and relationship between hormones and brain, 11, 22
risks and benefits of, 235, 247, 251, 255
estrogen/testosterone combinations, 112–15
heart disease and, 117
hormone modulation and, 189
nutrition and, 200
types of, 113–15
exercise, 242
AD and, 152–53
hormone modulation and, 191–93, 199–202, 205
PMS and, 145

falling, 162–63, 165, 170–71
fibromyalgia, 31–32, 220, 230–31, 247
Fink, George, 42
fluoxetine, *see* Prozac
follicle stimulating hormone (FSH), 97, 111, 121, 184, 191, 202, 211, 246
cyclical changes in, 60
hormone modulation and, 215, 219, 221, 226
memory and, 79, 83
mood problems and, 32, 50, 56, 59–61
Food and Drug Administration (FDA), 113, 120, 174, 182–83, 187
AD and, 159–60, 166
forgetfulness, *see* memory, memory loss
Fosamax (Alendronate), 244, 246, 253
Frank, Robert T., 129
frontal lobe, 76, 154

gall bladder disease, 248
gama-aminobutyric acid (GABA), 16, 48
Gapstur, Susan, 235
Gelfand, Morrie, 109
ginkgo biloba, 12, 84, 204
goiters, 52

Grady, Denise, 174
Graves' disease, 52, 152–53
Grodstein, Francine, 232–34, 237, 242
Grossman, Steven, 267

Haggerty, J. J., Jr., 53
Hamilton Scale of Depression, 38–39, 44
Hashimoto's thyroiditis, 26–27, 53, 86–87, 140, 212, 214, 219
heart disease, 5–7
 AD and, 152
 diet and exercise for, 199–201
 estrogen and, 5, 7, 116–17, 144, 164, 193, 196–97, 239–42, 253, 265
 HRT and, 118, 175–78, 187, 193–99, 202, 204, 208, 230, 233–34, 239–44, 249, 252–54, 257, 260
 mortalities from, 239–41, 257
 prevention of, 208
 testosterone and, 100–101, 116–18, 223
Henderson, Victor W., 154–55
herbs, 121, 201, 203–5
Herrington, David, 196
HERS study, 194–98
hip fractures, 243–44
hippocampus, 75–76, 87, 154
homeopathy, 201, 205
Honig, Peter, 174–75
Honjo, H., 157
hormonal histories, 210–12
hormone levels, normalization of, 64
hormone modulation, 2–8, 94–98, 173–257
 in action, 191–93
 AD and, 169, 175, 194, 198, 202, 204, 208, 213–14, 271, 274
 alternative medicine and, 201–4, 206
 as choice for you and your physician, 227–28
 cognition and, 70–71, 73, 94–96, 175, 183, 187–88, 202, 204, 209–10, 213–14, 216–17, 223–24, 228, 259, 261, 264, 267, 270–71, 274–75
 complementary medicine and, 204–5
 diagnosis and treatment in, 209–28
 energy and, 121, 177, 189, 191–92, 206, 212, 216–17, 220–23
 exercise and, 191–93, 199–202, 205
 future of, 269–75
 heart disease and, 7, 175–78, 187, 193–99, 202, 204, 208, 223
 hormonal histories in, 210–12
 hormonal lab testing in, 213–15
 how to get it, 259–67
 individualization and flexibility of, 63–64, 96–97, 113–14, 173–76, 178, 190, 208–10, 214, 227–29, 263–64, 274
 memory and, 4, 73, 80, 82, 84, 95, 175, 204, 206, 208–10, 212–14, 216, 221–22, 224–25, 259–61, 263–64, 267, 270–71, 274
 menopause and, 94–98, 173–209, 211–12, 214–16, 219, 222–26, 259–61, 266–67, 269–72, 274
 mental sharpness and, 73, 90, 93–94, 98, 176, 188, 204, 231, 270
 mood disorders and, 4, 28, 47–64, 94–98, 175–77, 181, 183, 185–88, 190–92, 201–6, 208–14, 216–28, 259–61, 263–64, 266–67, 269–72, 274
 naturalness of, 205–7
 origins of, 3–5
 physical exams in, 212–13
 PMS and, 130, 132, 139, 143, 175, 181, 190, 209–11, 213–14, 218, 223, 226–27
 promise of, 7–8

sexual health and, 97, 122, 127, 175, 177, 181, 184–85, 188, 202, 206–14, 221–25, 227–28, 263–64, 266–67, 270–71, 274–75
types of estrogen in, 178–81
types of progesterone in, 179–84
weighing risks and benefits of, 229–57, 259–61, 265–67, 272
Hormone of Desire, The (Rako), 99–100, 106–8, 114, 205, 266–67
hormone replacement therapy (HRT), 121
AD and, 161, 164, 168
for cognition, 71–75
exercise and, 200–201
going beyond standard doses of, 64, 175–78
heart disease and, 118, 230, 233–34, 239–44, 249, 252–54, 257, 260, 265
hormone modulation and, 173, 175–89, 193–99, 201–2, 204, 206–8, 214, 218, 259–60, 265, 269–70, 272
introducing progestins into, 186–88
memory lapses and, 71–72, 74, 231, 249–50, 252, 255–56
mood problems and, 40–41, 48–50, 52, 63–64, 229, 231, 246, 249–50, 252, 254, 256
as natural, 5–6
osteoarthritis and, 247
PMS and, 143, 252
sexual health and, 99, 118, 229, 231, 245–46, 250–52, 254, 256
weighing risks and benefits of, 229–57, 259–61, 265–67, 272
hormones:
as natural medicine, 205–7
principles of healing with, 93–98
rebalancing of, 4
Hormones (Love), 257
hot flashes, 12, 22–24, 32, 46, 56, 69, 74–75, 83, 94–97, 111, 121, 161, 244–45, 260, 266, 274
estrogen and, 176, 179, 187–89, 211–12, 224–25, 227, 249
HRT and, 175, 181, 201–2, 222, 230, 254
HRT, *see* hormone replacement therapy
hyperplasia, 186–87, 190
hyperthyroidism:
AD and, 152
diagnostic blood testing for, 214
mood problems and, 52–54, 58
hypothyroidism, 4, 26, 169
diagnostic blood testing for, 214
hormone modulation and, 214, 220
and HRT risks and benefits, 250
memory and, 86–87
mental sharpness and, 89
mood problems and, 43, 52–54, 56, 58, 62

IBS (irritable bowel syndrome), 230, 247
imipramine, 55, 203
insomnia, 247
hormone modulation and, 212, 217–18, 222, 224–27, 272
and HRT risks and benefits, 230, 256
PMS and, 130
Ireland, Patricia, 8
irritable bowel syndrome (IBS), 230, 247

Jordan, V. Craig, 252–53
Journal of American Geriatrics Society, 243
Journal of Clinical Endocrinology and Metabolism, 265
Journal of Neurology, 149
Journal of the American Medical Association, 21, 194–95, 235, 252, 261
Journal Watch, 236

Klonopin, 139
Kobayashi, Yutaka, 18
Kupfer, David, 42–43

Lawrence, Marcia, 94, 120
learning, 25, 27
Leisure World study, 155–56
LH, *see* luteinizing hormone
life expectancies, 5–6, 205
light therapy, 145
lithium, 54–55, 111, 266
Love, Susan, 257
Lupron, 133, 143–44
luteinizing hormone (LH), 118, 202, 211
  cyclical changes in, 60
  hormone modulation and, 215, 226
  mood problems and, 59–61
  sexual health and, 106, 124–25

McGwire, Mark, 120
macular degeneration, 7, 230–31, 248
magnesium, 144
malapropisms, 69, 71, 88, 231
mammillary bodies, 75–76
manic behavior, 55, 68
manic-depression, 18, 27, 29, 41, 53–56, 94, 134, 266
Maturitas, 184
medical cross-talk, 265–67, 273
medoxyprogesterone (Provera), 181, 183, 188, 190, 194–96, 198, 242
memory, memory loss, 1–6, 8, 35, 70–71, 74–88
  AD and, 75, 77, 84–86, 147–48, 151, 154, 158–59, 161–62, 164–67, 169–70
  brain and, 9–14, 18, 20–22, 24–29, 75–78, 85, 87
  diagnostic blood tests for, 213–14
  estrogen and, 71–72, 73, 74–87, 90, 95, 206, 216, 221, 224–25, 249–50, 255–56, 259
  ginkgo biloba and, 204
  hormone modulation and, 4, 73, 84, 95, 208–10, 212–14, 221, 224–25, 252, 261, 263–64, 267, 270–71, 274
  HRT and, 71–72, 74, 175, 231
  mechanisms of, 75–78
  severity of, 83
  testosterone and, 81–84, 87, 101, 221, 224–25
  thyroid hormone and, 83–88
  verbal, 80–82, 90
  visual, 80, 82
  what to do about, 82–84
men, 3–4
  AD in, 149–51
  cognition in, 65, 72
  depression in, 34, 44, 65–68, 112, 270
  Hashimoto's thyroiditis in, 86
  heart disease in, 117–18
  on high doses of MT, 251
  hormone modulation for, 4, 127
  sexual health of, 65, 104, 123–27
menopause, 1–2, 4–6, 8, 69, 111–12, 120–21, 143
  AD and, 147, 149, 151, 154–56, 161–64, 166, 171
  andropause vs., 123
  cognition and, 70–71, 73
  estrogen and, 2, 14, 21–23, 28, 242–47, 255, 266
  heart disease and, 116–17, 257
  hormone modulation and, 94–98, 173–208, 209, 216, 219, 222–26, 259–61, 266–67, 269–72, 274
  HRT and, 230, 234–35, 238–39, 259–61
  memory and, 78–81, 83, 85–87
  mental sharpness and, 88–89, 91–92
  mood problems and, 32, 35–36, 40, 43, 47, 49–51, 56, 58, 61, 65, 79
  PMS and, 134–35
  premature, 222–23, 250
  sexual health and, 79, 100–101, 103–5, 109, 112, 122

surgical, 223–25, 271
symptoms of, 9–12, 73, 173–77, 184, 187–88, 201–2, 205–8, 211–12, 224, 222, 234, 244–45, 255, 260, 266, 274
testosterone and, 2, 114–15, 117–18, 246, 267
thyroid hormone and, 56, 250
Menopause and Madness (Lawrence), 94, 120
menstrual cycle:
hormone level changes in, 59–60
hormone modulation and, 210–11, 213–14
mental sharpness, 1, 4–5, 69–71, 88–93, 208
AD and, 161–62
hormone modulation and, 73, 90, 93–94, 98, 176, 188, 204, 231, 270
mesterolone, 26, 66, 112
methyltestosterone (MT), 32, 95, 103, 113–17, 251, 267
energy and, 121
heart disease and, 117
hormone modulation and, 183–84, 191, 217–18, 221, 223, 225, 227
sexual health and, 111
Miacalcin (calcitonin), 244
migraine headaches, 45, 230
hormone modulation and, 220–22
mind-body medicine, 144
minerals, 144, 205
monoamine oxidase (MAO), 3–4
AD and, 158–59, 161
hormone modulation and, 185–86, 211, 213–15, 219, 222, 225–26
memory and, 77
mood problems and, 18–19, 32, 37, 39, 42, 45, 48, 59–62, 65–66
PMS and, 137, 142
monoamine oxidase (MAO) inhibitors, 12, 161–62, 166–67, 262
mood states, mood disorders, 1–8, 31–69, 74, 83
AD and, 149–50, 159, 167
brain and, 9–29, 41–42, 45, 48, 68, 113
and diet and exercise, 199–200
estrogen and, 32–52, 56–61, 63–67, 77, 82, 93, 95–96, 98, 111–13, 216–18, 220, 246, 249–50, 256
ginkgo biloba for, 204
hormone evaluations for, 68
hormone modulation and, 4, 28, 47–64, 94–98, 190–92, 201–6, 208–14, 216–28, 259–61, 263–64, 266–67, 269–72, 274
HRT and, 40–41, 48–50, 52, 63–64, 175–77, 181, 183, 185–88, 208, 229–231, 252, 254, 256
menopause and, 32, 35–36, 40, 43, 47, 49–51, 56, 58, 61, 65, 79
molecular drama of, 15–22
neurotransmitters and, 18, 33, 37, 41–43, 48, 65, 77
perimenopause and, 32–33, 35–36, 40–41, 43, 46–47, 50–52, 58, 61
PMS and, 34, 58–62, 129–38, 140–45
progesterone and, 35, 40–41, 46–51, 55, 59–61, 63, 95–97, 111, 119, 121
St. John's Wort for, 203–5
testosterone and, 32–33, 40–41, 43–44, 46, 48, 51–52, 58–61, 64–67, 95–96, 100, 111–16, 126, 246, 256
thyroid hormone and, 43, 52–64
see also specific disorders
Morrow, Monica, 235
MT, *see* methyltestosterone

Nachtigall, Lila E., 249
Naftolin, Frederick, 259
National Cancer Institute, 234, 236

National Institute of Mental Health (NIMH), 4, 20, 36, 135, 203
National Institutes of Health (NIH), 19, 35, 37–38, 133–34, 167, 185, 200, 271
National Institutes of Health Sciences, 202
natural treatments, 84
Neurology, 154
neurons:
AD and, 152–54, 157–58, 166–67, 169
in cognition and memory, 76
mood problems and, 41
sexual health and, 113
neurotransmission, neurotransmitters, 15–20, 24, 200, 275
AD and, 154, 158–59, 166, 169
hormone modulation and, 185, 203, 219, 270
interaction between receptors and, 16–17, 19–20, 22, 24, 26, 33, 42–43, 45
memory and, 75–77
mood problems and, 18, 33, 37, 41–43, 65
PMS and, 134–36, 140–43
and relationship between hormones and brain, 12, 15–20, 22, 24–26
testosterone and, 101
thyroid hormone and, 54
see also specific neurotransmitters
New England Journal of Medicine, 115, 198, 200, 232–33, 240
new menopausal woman, 207–8
New York Times, 42, 101, 174, 234–35
night sweats, 23–24, 32, 46, 56, 69, 74–75, 83, 94–97, 161, 184, 211–12, 244–45, 266, 274
estrogen and, 176–77, 179, 187, 225, 249
HRT and, 181, 227, 230
NIH (National Institutes of Health), 19, 35, 37–38, 133–34, 167, 185, 200, 271
NIMH (National Institute of Mental Health), 4, 20, 36, 135, 203
nipples, 106–7
norepinephrine, 158, 203
mood problems and, 16, 18–19, 33, 37, 41–43, 48, 65, 77
PMS and, 134, 141–42, 145
thyroid hormone and, 54
norethindrone, 134, 181, 183
norgestimate, 180–81
Nurses Health Study, 198–99
nutrition, 198–202, 242
AD and, 152–53, 164–65
hormone modulation and, 192–93, 199–202, 205
PMS and, 144–45
sexual health and, 122–23

O'Brien, William, 40
obsessive compulsive disorder (OCD), 110
Ohkura, T., 156–57
oral contraceptives, 46–49, 118, 248
cognition and, 97–98
mood problems and, 46–47, 110, 177
PMS and, 134–35, 142, 144
progesterone administered with, 49
sexual health and, 106, 108
Oregon Regional Primate Center, 197
Ortho-Prefest, 180–81
osteoarthritis, 247
osteoporosis, 6–7, 86, 170–71, 214
diet and exercise for, 199–201
estrogen and, 144, 164, 171, 242–46
HRT and, 175, 177–78, 187, 194, 198, 202, 204, 208, 230–31, 234, 239, 242–46, 249–50, 254, 260
ovaries, 117–19
AD and, 152
assessing function of, 59, 272
cancer of, 192–93, 212, 214, 224–25, 245

estrogen and, 187, 191, 224–25
heart disease and, 117
hormone modulation and, 191–93, 224–25
HRT and, 205, 250, 256
PMS and, 134–35, 140–42
sexual health and, 100, 103–4, 106, 109, 115
surgical removal of, 224–25, 271
testosterone and, 100, 103–4, 191, 224–25
ovulation, blocking of, 143–44

Paganini-Hill, A., 155–56
panic attacks, 9–10
AD and, 161
mood problems and, 47
PMS and, 137–40
paranoia, 165, 170–71
Parkinson's disease, 13, 271
Pause, The (Barbach), 36–37
Paxil:
mood problems and, 12, 18, 56, 216, 218, 226
PMS and, 135–36, 138
perceptual-restructuring, 72–73
perimenopause, 11–13, 28–29, 199, 274
cognition and, 73
estrogen and, 121, 214
hormone modulation and, 96, 98, 174, 191–92, 208, 209, 211, 220–22, 231, 239, 264, 269
memory and, 83
mental sharpness and, 91, 93
mood problems and, 32–33, 35–36, 40–41, 43, 46–47, 50–52, 58, 61, 226
PMS and, 143
sexual health and, 100, 102, 106, 110
symptoms of, 83, 211–12
Persky, Harold, 102–3
PET (proton emission tomography) scans, 20–21
Pfaff, Donald, 14
physical exams, 212–13
physicians, 227–28
communicating with, 261–65, 272–74
communication between, 265–67, 272–74
as medical facilitators, 262–63
Physician's Desk Reference (PDR), The, 174
phytoestrogens, 201–2, 206
Pilette, Wilfrid, 226
PMS, *see* premenstrual tension syndrome
Postmenopausal Estrogen/Progestin Interventions (PEPI) Trial, 197
postmenopause:
AD and, 156, 158, 168
estrogen and, 237–38, 248
heart disease and, 117, 195, 239–40
hormone modulation and, 21, 29, 96, 174, 208, 215, 239, 264
HRT and, 177, 181, 195, 197–99, 201–2, 208, 231–32, 248
memory and, 78
mental sharpness and, 91, 93
mood problems and, 38, 40–41, 43, 46, 50
PMS and, 143
sexual health and, 13, 100, 102, 106, 109–10
postpartum period:
depression and, 28, 34–35, 43–45, 53, 56, 58, 61–63, 106, 110–11, 220–21
psychoses in, 35
sexual health and, 106
Prange, A. J., Jr., 53
Premarin, 21, 50, 64, 77, 113
AD and, 149, 161
in hormone modulation, 178–81, 217, 221–22
in HRT, 242, 249, 256

premenopause:
hormone modulation and, 197, 206, 226, 228
mood problems and, 38
PMS and, 134, 137, 143
premenstrual phase:
mood problems and, 32, 34, 43
secretory phase and, 59
premenstrual tension syndrome (PMS), 3, 5, 28, 50, 110, 129–46, 245
causes of, 132–35, 142–43, 145
cognition and, 73
estrogen and, 130–46
hormone modulation and, 130, 132, 139, 143, 175, 181, 190, 209–11, 213–14, 218, 223, 226–27
HRT and, 143, 252
mood problems and, 34, 58–62, 129–38, 140–45
oral contraceptives and, 134–35, 142, 144
progesterone and, 131–35, 137–38, 140, 142–44
seasonal changes in, 145
serotonin and, 132, 134–37, 140–42
sexual health and, 106
symptoms of, 62, 129–31, 133, 136–45
thyroid hormone and, 53, 55–56, 131
treatments for, 132, 135–46
women's vulnerability to, 133–35
Prempro, 49, 118, 180, 198, 242
progesterone, 1, 7, 28, 74
AD and, 161
administration of, 24–25, 47, 49–51, 121, 176, 179–84, 187–90, 197, 252
brain activity and, 21, 27
breast cancer and, 236–37
cognition and, 24–25, 72–73, 94–98, 119
counteraction of testosterone and, 118–19
cyclical changes in, 60
energy and, 121
hormone modulation and, 11–12, 173–74, 176, 179–90, 194–99, 205–6, 211, 215, 222–23, 227, 264, 272
HRT and, 173, 176, 181–83, 186–88, 229, 231, 251–52
mental sharpness and, 92–94
micronized, 119, 182–84, 186, 189–90, 194, 197, 222–23, 227
mood problems and, 24–25, 35, 40–41, 46–51, 55, 59–61, 63, 95–97, 111, 119, 121
PMS and, 131–35, 137–38, 140, 142–44
problems and concerns about, 48, 50, 184–90
search for heart-healthy, 197–99
sexual health and, 99, 109–11, 118–19, 222–23
synthetic (progestin), 181–83, 185–88, 190, 194–97, 248, 249, 251–52, 259
types of, 179–84
uterine cancer and, 24–25, 98, 119, 121, 176, 179–81, 231, 251–52, 269
see also estrogen/progesterone combinations
progestin (synthetic progesterone), 181–83, 185–88, 190, 194–97, 248, 249, 251–52, 259
Prometrium, 182
proprioception, 170
prostate cancer, 125
proton emission tomography (PET) scans, 20–21
Provera (medoxyprogesterone), 181, 183, 188, 190, 194–96, 198, 242
Prozac (fluoxetine), 3, 203
hormone modulation and, 209

mood problems and, 12, 18, 22–23, 31, 34, 42–44, 47, 56, 63, 66, 74, 98, 111, 131, 216, 218, 226, 269–70
PMS and, 131, 135–36
side effects of, 66
psychiatric status, markers of, 59
psychoneuroendocrinology, 1–2, 28, 261
psychoses, 28, 35, 68, 120, 170
mood problems and, 38–39
pyridoxine (vitamin B_6), 12, 144

Rako, Susan, 205–6, 266–67
on testosterone, 99–100, 106–8, 114, 120
raloxifene, 206, 252–54
Research Initiatives Committee, 271
Resnick, Susan, 82

SADS (seasonal affective disorder syndrome), 145, 226
St. John's Wort, 12, 203–5
Sarrel, Phillip, 109–10
Schairer, Catherine, 234
schizophrenia, 28, 42, 55, 68
Schmidt, Peter, 135
Screaming to Be Heard (Vliet), 132
seasonal affective disorder syndrome (SADS), 145, 226
selective estrogen receptor modulators (SERMs), 252–54
selective serotonin reuptake inhibitors (SSRIs), 203–4
mood problems and, 12–13, 18, 22–23, 62, 216, 226
PMS and, 131, 136–39, 141–43
selegiline, 166–67
Sellers, Thomas, 235
serotonin, 16, 18–19, 24
AD and, 158
hormone modulation and, 203–5, 220
memory and, 77
mood problems and, 22–23, 37, 41–43, 45, 48, 65
PMS and, 132, 134–37, 140–42
thyroid hormone and, 54
sertraline, *see* Zoloft
sex, sexuality, 1–7, 35, 37, 83, 89, 99–127, 143, 201, 260
brain and, 10–14, 22–29, 104–5
estrogen and, 51–52, 80, 101–12, 118, 122, 124
hormone modulation and, 97, 122, 127, 175, 177, 181, 184–85, 188, 202, 206–14, 221–25, 227–28, 263–64, 266–67, 270–71, 274–75
HRT and, 99, 118, 229, 231, 245–46, 250–52, 254, 256
of men, 65, 104, 123–27
menopause and, 79, 100–101, 103–5, 109, 112, 122
mesterolone and, 66
mood problems and, 31–32, 48, 111–12
progesterone and, 99, 109–11, 118–19
testosterone and, 51–52, 61, 99–112, 114–16, 120, 122, 124–25
thyroid hormone and, 53, 122
whole-person approach to, 122–23
sex hormone binding globulin (SHBG), 104–5, 124
Shaywitz, Sally, 21, 77
Sherwin, Barbara, 22, 40–41, 50, 52, 149, 186
cholesterol studies of, 117
on sexual health, 108–9, 112
verbal memory studies of, 80–82, 90
Sichel, Deborah, 68
sildenafil citrate (Viagra), 13, 127
Simon, James, 259
smoking, 198–99, 212, 248
AD and, 152–53, 161
heart disease and, 240–42

soy foods and products, 201–2, 206
specialists:
  bringing together, 261–62
  communication between, 265–67, 273
  hormone modulation and, 261–62, 264–67
SSRIs, *see* selective serotonin reuptake inhibitors
Stahl, Stephen M., 270
stress:
  hormone modulation and, 193, 205
  mood problems and, 58
  PMS and, 130, 134, 141
  sexual health and, 106, 122–23
  testosterone and, 112
stroke, 7, 230–31, 233
Studd, John, 138
synapses, 16–18, 24, 76, 136

tacrine (Cognex), 13, 159–60, 166, 271
"Tad of Testosterone Adds Zest to Menopause, A" (Brody), 101–2
tamoxifen, 206, 252–53
testosterone, 1–4, 37
  administration of, 113–15, 125–26
  anabolic effects of, 122
  brain and, 11–12, 15, 18, 21–28, 116, 126
  cognition and, 72–73, 95–96, 126
  counteraction of progesterone and, 118–19
  DHEA and, 119
  energy and, 120–22
  estrogen/progesterone combinations and, 118
  facts about, 113–20
  free, 105, 109, 111, 114–15, 117, 124, 184, 191, 211, 213, 215
  heart disease and, 100–101, 116–18
  hormone modulation and, 173, 181, 184–86, 189, 191–92, 205–6, 211, 213, 215, 217–18, 221–25, 227–28, 246–47, 250–51, 256, 264, 266–67, 270
  memory and, 81–84, 87, 101
  for men, 65–67, 124–27
  menopause and, 2, 114–15, 117–18
  mental sharpness and, 89–90
  mood problems and, 32–33, 40–41, 43–44, 46, 48, 51–52, 58–61, 64–67, 100, 111–16, 126
  nutrition and, 200
  PMS and, 138
  sexual health and, 51–52, 61, 99–112, 114–16, 120, 122, 124–25
  side effects of, 101–3, 113, 115–16
  types of, 113–14
  see also estrogen/testosterone combinations
testosterone cypionate (Depo-Testosterone), 114, 126
testosterone replacement therapy, 103, 108–10
thrombophlebitis, 229–30, 248
thyroid function, assessment of, 59
thyroid hormone, 1, 3–4
  brain and, 12
  cognition and, 73
  diagnostic blood testing for, 214–15
  hormone modulation and, 211–15, 218–20
  memory and, 26–27, 83–88
  mental sharpness and, 26–27, 89
  mood problems and, 26–27, 43, 52–64
  PMS and, 53, 55–56, 131
  risks of, 250
  sexual health and, 53, 122
thyroid-stimulating hormone (TSH):
  hormone modulation and, 211, 213–15, 219
  mood problems and, 53–54, 56
  PMS and, 140
thyroxine (T4), 4, 26
  mood problems and, 53–54, 56, 59
  PMS and, 139–40

Trends in the Incidence of Coronary Heart Disease and Changes in Lifestyle and Diet in Women, 198
tricyclic antidepressants, 12, 55, 161, 203, 272–73
triiodothyronine (T3), 53, 59, 211, 213–15

ultrasound, 190, 213–14, 223, 225
uterine cancer, 11
  estrogen and, 40, 47–48, 50, 57, 179–81, 197–98, 212, 251, 269
  hormone modulation and, 176, 183–84, 208
  HRT and, 176, 179–81, 186–87, 201, 251–54, 260
  progesterone and, 24–25, 98, 119, 121, 176, 179–81, 197, 231, 251–52

vagina, 14
  administering hormones through, 24–25, 47, 50–51, 119, 121, 179–84, 190, 197, 223
  dryness of, 12, 22, 32, 69, 83, 100–102, 105–6, 122, 179, 211, 244–45, 256, 274
  hormone modulation and, 202, 213
  sexual health and, 100–102, 105–6, 122
  thinning of wall of, 105–6, 122
verbal memory, 80–82, 90
Viagra (sildenafil citrate), 13, 127
visual (nonverbal) memory, 80, 82
visual problems, 221
vitamins:
  B, 12, 144
  B_6 (pyridoxine), 12, 144
  B_{12}, 169
  C, 157
  E (alpha tocopheral), 157, 166–67
  hormone modulation and, 205
  PMS and, 144
Vivelle, 121, 217
Vliet, Elizabeth, 132
Vogel, William, 3–4, 15, 73, 266
  cognition and, 72
  mental sharpness and, 91–92
  mood problems and, 37–38

Wellbutrin, 23, 56
Women's Health Initiative Study (WHIS), 167–69, 194, 197–98, 237
Women's Moods (Sichel), 68
Worcester Foundation, 4, 18, 252
Worcester State Hospital, 3–4, 15, 38–39, 112, 216

Xanax, 137–38, 161

Yaffe, K. G., 156

Zoloft (sertraline), 203
  mood problems and, 12, 18, 45, 56, 63, 66, 216, 218
  PMS and, 135–36, 142–43
Zweifel, Julianne, 40